Energy Law in Romania

Energy Law in Romania

Second Edition

Cătălin Gabriel Stănescu

This book was originally published as a monograph in the International Encyclopaedia of Laws/Energy Law.

Founding Editor: Roger Blanpain
General Editor: Frank Hendrickx
Volume Editors: Kurt Deketelaere & Bram Delvaux

Wolters Kluwer

Published by:
Kluwer Law International B.V.
PO Box 316
2400 AH Alphen aan den Rijn
The Netherlands
E-mail: international-sales@wolterskluwer.com
Website: lrus.wolterskluwer.com

Sold and distributed in North, Central and South America by:
Wolters Kluwer Legal & Regulatory U.S.
7201 McKinney Circle
Frederick, MD 21704
United States of America
Email: customer.service@wolterskluwer.com

Sold and distributed in all other countries by:
Air Business Subscriptions
Rockwood House
Haywards Heath
West Sussex
RH16 3DH
United Kingdom
Email: international-customerservice@wolterskluwer.com

Printed on acid-free paper

ISBN 978-94-035-0593-0

e-Book: ISBN 978-94-035-0222-9
web-PDF: ISBN 978-94-0350-230-4

Printed in the United Kingdom.

To Natochka

The Author

Cătălin Gabriel Stănescu (born on 9 September 1980) studied law at Al.I.Cuza University, Faculty of Law, in Iasi, Romania (1999–2003). He obtained an MA in political studies in 2006 from the National School of Political and Administrative Studies, in Bucharest, Romania, and an LLM in international business law in 2011 from Central European University, Budapest, Hungary. In 2015, he graduated *summa cum laude* from the doctoral programme of Central European University, with a thesis dedicated to self-help and private debt collection, which was published the same year under the title 'Self-Help, Private Debt Collection and the Concomitant Risks: A Comparative Law Analysis.'

He has extensive practical experience in the field of private Romanian law. In 2004 he started working as a legal counsel for Ploiesti City Hall, passed the Bar exam in 2005 and opened his own practice in 2006. He also worked as an in-house counsel for the Romanian subsidiaries of two oil and gas multinationals, specializing in upstream oil and gas consultancy. Currently he holds a postdoctoral position at the Center for Enterprise Liability, University of Copenhagen, Denmark dealing with long-term energy contracts.

He has several publications, both in English and Romanian, dedicated to topics such as: self-help repossession, piercing the corporate veil, fraudulent transfers, oil and gas contracts and renewable energy, all analysed from a comparative perspective.

Table of Contents

3

The Author 5

List of Abbreviations 19

Acknowledgements 21

General Introduction 23

Chapter 1. Background 23

§1. Geography 23

§2. Population 23

§3. Political System 24

§4. Population and Employment Statistics 28

§5. Social and Cultural Values 34

Chapter 2. The Economic System 37

Chapter 3. The Legal System 38

§1. Legal Family or Tradition 38

§2. Relation and Hierarchy Between the Sources of Law 39

Part I. Regulatory Framework Concerning Electricity 43

Chapter 1. General Introduction 43

§1. Principles Governing Electricity Sector 45

§2. Fundamental Purposes 45

§3. The Energy Strategy and Policy: The Energy Programme 46

Chapter 2. Authorities and Competences 49

§1. The Relevant Ministry: Duties 49

§2. Regulatory Authority 49

Chapter 3. Authorizations and Licenses 53

§1. Activities Requiring Authorization and License 53

§2. Authorization and Licensing Procedures 54

§3. Suspension and Withdrawal of Authorizations and Licenses 55

§4. The Challenging Procedure 55

§5. The Rights and Obligations Arising from Authorizations and Licenses 56

§6. Obligations for Holders of Authorizations and Licenses 57

§7. Rights and Obligations of Authorization and License Holders over the Property of Third Parties 58

§8. The Right to Request Expropriation 59

§9. Access to Electrical Network: Exception 59

Chapter 4. Production 60

§1. Procedures for New Power Plants: Public Tender 60

§2. The Concession Agreement: Object and Granting Authority 60

§3. The Obligations of the Producers 61

§4. The Rights of the Producers 61

§5. Electricity Produced from Renewable Energy Sources and in Highly Efficient Cogeneration 62

§6. Electricity Produced from Renewable Energy Sources. General Framework 64

I. Beneficiaries of the Support Scheme 65

II. National Target Levels Concerning the Percentage of Electricity Produced from Renewable Energy Sources 65

Chapter 5. Transmission of Electricity 67

§1. The Certification of the Transmission and System Operator 67

§2. The Certification of the Transmission and System Operator Controlled by Third Party Countries 67

§3. The Re-evaluation of Certification of the Transmission and System Operator 68

§4. The Obligation of the Transmission and System Operator 68

§5. Duties of the Owner of a Transmission Network 71

§6. The Independence of the Owner of a Transmission Network 71

§7. Interdictions Regarding the Transmission Network 72

Chapter 6. Electricity Distribution 73

§1. Distribution Operator 73

§2. The Obligations of the Owner of a Distribution Network 74

§3. Separation Between Distribution and Supply Activities 75

§4. Measures for Protecting the Distribution Networks 76

§5. Isolated Distribution System 77

§6. Electrification of Localities and Expansion of Distribution Networks 77

Chapter 7. The Electricity Market 78

§1. The Participants in the Electricity Market: General Obligations 78

§2. Electricity Market Operator 79

§3. The Operation of the Electricity Regulated Market 79

§4. The Role of NRAE in the Operation of the Regulated Electricity Market 80

§5. The Operation of the Competitive Market 80

§6. The Role of NRAE in the Operation of the Competitive Market 81

§7. Limitation of Prices and Tariffs 81

Chapter 8. The Green Certificates Market 82

§1. Role of the Market Operator (OPCOM) in Administering the Green Certificates Market 82

§2. The Value of Green Certificates 83

§3. The Bilateral Contracts Market of Green Certificates 84
I. The Functioning of the Centralized Market for Bilateral Contracts Concluded as a Result of a Public Tender 84
II. Rights and Obligations of the Seller 85
III. Rights and Obligations of the Buyer 85
IV. The Functioning of the Bilateral Contracts Market Concluded Through Direct Negotiation 86

§4. The Functioning of the Centralized Green Certificates Market 86

§5. Sanctions 87

§6. Resolution of Disputes Arising on the Green Certificates Market 88

Chapter 9. Electricity Supply 89

§1. Direct Electricity Lines 89

§2. Connection Agreement 89

§3. Last Instance Suppliers: Rights and Obligations 90

§4. Contracting Obligation: Termination of the Supply Agreement 91

§5. Liabilities of the Supplier 91

§6. Liabilities of the Final Customers 92

§7. The Rights of the Final Customers 92

§8. Protection of Vulnerable Customers 93

§9. Measurement of Electricity 93

Part II. Regulatory Framework Concerning Gas 95

Chapter 1. General Introduction 96

§1. Energy Strategy and Policy in Natural Gas Sector 96

§2. The Relevant Ministry: Competences 97

§3. NRAE: Competences Concerning the Natural Gas Sector 98

§4. Regulatory Body: Power to Conduct Investigations 98

§5. General Framework of Concessions Concerning Natural Gas Sector 99

Chapter 2. Licensing and Authorization Regime 100

§1. Procedures, Terms and Conditions for Granting Authorizations and Licenses 100

§2. Common and Specific Documents Required for All Types of Authorizations and Licenses 101

§3. Approval or Refusal of Request for Granting Authorization or License 102

§4. The Procedure for Modifying Authorizations and Licenses 103

§5. The Procedure for Suspending or Withdrawing Authorizations or Licenses 104

Chapter 3. Rights and Obligations of License Holders in the Natural Gas Sector 105

§1. Common Rights and Obligations of License Holders in the Natural Gas Sector 105

§2. Rights and Obligations of the Grantees Concerning Goods Belonging to Third Parties 105

§3. Right of Use and Easement Right: Other Rights 106

Chapter 4. Production and Exploitation 108

Chapter 5. Transportation 109

§1. The National Transport System: Certification of the System Operator 109

§2. Rights and Obligations of the System Operator 110

§3. Duties of the System User 111

§4. Duties of the Owner of the Transport System 111

Chapter 6. Distribution 113

§1. Right and Obligations of the Distribution Operator 115

§2. Closed Distribution System 115

Chapter 7. The Natural Gas Market 117

§1. The Functioning of the Regulated Natural Gas Market 117

§2. The Functioning of the Competitive Market 118

§3. Prices and Tariffs 119

Chapter 8. Storage, Supply and Consumption 120

§1. Rights and Duties of the Storage Operator 120

§2. Rights and Duties of the Natural Gas Supplier 121

§3. Duties of the Supplier of Last Resort 122

§4. Rights and Duties of Final Consumers of Natural Gas 123

§5. Evolution of Production and Consumption 124

Part III. Regulatory Framework Concerning Petroleum 127

Chapter 1. General Introduction 128

Chapter 2. Production 130

Chapter 3. Concession Agreements 132

§1. General Regime of Concession Agreements 132

§2. Special Regime of Concession Agreements under the Petroleum Law 135
I. Prospection Works 136
II. Exploration Works 136
III. The Concession Agreement: Initiation 137
IV. The Concession Agreement: The Opening, Evaluation and Negotiation Committees 139
V. The Concession Agreement: The Selection Criteria and the Corresponding Scores 140
VI. The Concession Agreement: The Opening of the Bids 140
VII. The Concession Agreement: Challenging the Public Call for Tenders 140
VIII. Termination of Concession Agreements and Its Effects 141

Chapter 4. The Farm-Out Agreement (Transfer of Rights and Obligations Derived from Petroleum Agreements) 143

§1. Conditions of Validity 144
I. Conditions Concerning the Farmee 144
II. Conditions Concerning the Farmor 145
III. Conditions and Authorization Procedure 145

§2. The Object of the Transfer 147

§3. The Rights and Obligations of the Parties 148
I. Rights and Obligations of the Farmor 148
II. Rights and Obligations of the Farmee 149
III. Rights and Obligations of Third Parties: NAMR 150
IV. Rights and Obligations of Other Third Parties 151

§4. Termination of Farm-Out Agreements 151

Chapter 5. Exploitation 152

§1. Experimental Exploitation 152

§2. The Annual Programme 153

Chapter 6. Transportation 154

§1. The Transit of Oil 154

§2. Transportation of Oil through the National Transportation System 154

I. The Common Transporter 155

II. The Framework Agreement for Transportation of Crude Oil, Gasoline, Condensate and Liquid Ethane 156

III. Safety and Protection Areas 156

IV. Transportation Tariffs 157

Chapter 7. Distribution and Consumption 159

Part IV. Regulatory Framework Concerning Coal 161

Chapter 1. General Introduction 161

§1. General Regime for Coal 162

§2. Acquiring Rights of Usage and Access to the Lands on which Mining Activities are Performed 163

§3. Data and Information Concerning Mineral Resources/Reserves 164

§4. Termination of the Concession Agreement or of the Right to Administer 165

Chapter 2. Authorities and Competences 168

§1. Competences of the Relevant Minister 168

§2. Rights and Obligations of the Competent Authority 168

Chapter 3. Licenses and Permits: Right and Obligations. Other Duties 170

§1. Rights and Obligations of License or Permit Holders 170

§2. Transfer of a License for Concession 172

§3. Taxes, Royalties and Tariffs 174

§4. Suspension of the License or the Permit 175

§5. Cancellation of the License or Permit 176

§6. Authorizations 177

§7. Dispute Settlement and Arbitral Clauses 178

Chapter 4. Production 179

§1. The Prospection Permit 179

§2. The License for Exploration 180

Chapter 5. Exploitation 183

§1. License for Exploitation 183

§2. Permit for Exploitation 185

§3. Cessation of Mining Operations: Closing the Mines 186

Chapter 6. Distribution and Consumption 188

Part V. Interaction Between Energy Law and Environmental Law 189

Chapter 1. General Introduction 190

§1. Pollution Prevention 190

§2. General Principles of Environmental Protection 190

§3. Liability for Activities which Affect the Environment. Obligations of Natural and Legal Persons 191

I. Specific Obligations for Natural and Legal Persons Conducting Energy-Related Operations 192

II. General Regime of Liability: 'Polluter Must Pay' Principle 192

Chapter 2. Authorities with Competences in Environmental Protection: Powers and Duties 195

§1. Powers and Duties of the Central Public Authority 195

§2. Powers and Duties of Other Central and Local Authorities 196

§3. Taxes and Tariffs 197

Chapter 3. Authorization of Activities Having an Impact on the Environment 198

§1. ENVIRONMENTAL PERMIT AND AUTHORIZATION 198

§2. ENVIRONMENTAL IMPACT ASSESSMENT: STAGES – TYPES OF ASSESSMENT 199

Chapter 4. Areas of Protection 202

§1. REGULATION CONCERNING WASTE 202

§2. REGULATION OF AIR EMISSIONS: PROTECTION OF ATMOSPHERE AND CLIMATE CHANGE 202

§3. PROTECTION OF SOIL, SUBSOIL AND TERRESTRIAL ECOSYSTEMS 203

§4. PROTECTION OF FRESH WATER AND SEAWATER 203

§5. SANCTIONS AND REMEDIES 204

Part VI. Interaction Between Energy Law and Tax Law 205

Chapter 1. General Introduction 206

Chapter 2. Tax on Income or Profits Obtained in Romania 207

Chapter 3. Value Added Tax 210

Chapter 4. Excise Tax 212

Chapter 5. Special Taxes 216

§1. TAX ON DOMESTIC OIL 216

§2. SPECIAL BUILDINGS TAX OR 'THE POLE TAX' 216

§3. OTHER SPECIAL TAXES OUTSIDE THE FISCAL CODE: TAX ON NATURAL MONOPOLIES IN THE ELECTRICITY AND NATURAL GAS SECTOR 216

§4. THE QUOTA FOR DEVELOPMENT AND MODERNIZATION OF OIL AND GAS PRODUCTION 217

Chapter 6. Local Taxes 218

§1. Land Tax 218

§2. Tax for Issuance of Permits, Approvals or Authorizations 218

§3. Other Special Taxes 219

Part VII. Interaction Between Energy Law and Competition Law 221

Chapter 1. General Introduction 222

§1. Romanian Competition Law 223

§2. Main Violations 223

Chapter 2. The Competent Authority 225

§1. The Competition Council 225

§2. Main Powers and Duties 225

Chapter 3. The Electricity Sector and Electricity Market 227

§1. The Wholesale Electricity Market Operator (OPCOM) and Restriction of Competition 228

§2. The Regulated Retail Energy Market and Restriction to Competition 229

§3. The Electricity Market: Risks and Council's Recommendations 230

§4. Electricity Market: Concluded and Pending Investigations 232

Chapter 4. The Gas Market 234

Conclusions 237

Selected Bibliography 239

Index 241

List of Abbreviations

ACER	Agency for the Cooperation of Energy Regulators
ADDB	The Administration of Danube Delta Biosphere
ENG	Environment National Guard
GEO	Government Emergency Ordinance
GGEC	Greenhouse Gas Emissions Certificates
JOA	Joint Operations Agreement
NAEP	National Agency for Environmental Protection
NAMR	National Agency for Mineral Resources
NCC	New Civil Code
NES	National Energy System
NGF	National Geologic Fund
NRAE	National Regulatory Authority for Energy
OPCOM	The Romanian Electricity and Natural Gas Market Operator (in Romanian original *Operatorul pietei de energie electrica si de gaze naturale din Romania*)
RON	Romanian New Leu
TFEU	Treaty for the Functioning of the European Union
VAT	Value Added Tax

Acknowledgements

I would like to express my gratitude to the many people who saw me through the writing of this monograph by providing their support, talking things over, reading and offering comments or materials, editing and proofreading it.

I would like to thank my good friend, Bogdan Timofti, for ensuring access to his firm's database and materials that enabled me to write this monograph.

I would like to thank my good friend, Cosmin Andrei Artimof, for all his moral support and constant encouragement.

I would like to thank, Joeri Lauwers, for his understanding and constant support throughout the writing and the updating of this monograph, and all the crew of Wolters Kluwer who made the publishing of this monograph possible.

As always, a warm thank you goes to my mother for everything she has done for me.

General Introduction

Chapter 1. Background

§1. GEOGRAPHY

1. Romania is a medium-sized country located in the geographical centre of Europe (south-east of Central Europe), half distance between the Atlantic coast and the Ural Mountains, on the lower Danube (1,075 km) and with access to the Black Sea. It covers an area of 238,391 km^2 (92,043.2797 sq mi) and has 245 km (152,236 mi) of the Black Sea coastline. Its neighbours are: Hungary (west), Ukraine (north-west and north), Republic of Moldova (east), Ukraine (south-east), Bulgaria (south) and Serbia (south-west). According to the latest census, the stable population is about 19.04 million and some 79 people live on average per kilometre square (197 per sq mi). The capital city is Bucharest, with a population of 1.67 million people.

§2. POPULATION

2. Romania is composed of two main cultural-linguistic communities: one Romanian-speaking community representing the majority of the population (over 88.9% of the total population) and a small Hungarian-speaking community (6.5% of the total population). There are other minority ethnic groups such as Romas (3.3% of the total population), Ukrainians, Germans, Turks, Russians (all of them with less than 1% of the total population). The Romanian majority is widely spread throughout the country, while the Hungarian minority is mostly concentrated in the centre of Romania (Harghita and Covasna counties) and spread in the western parts. The official language of the country is Romanian, but minorities have the right to use their mother tongue in the administration where their presence is larger than 20%.

3. On March 2018, according to the results of the 2011 census, the population of Romania was as follows:

	October 2011	*December 2014*	*Increase/Decrease*	*Evolution*
Romania	20.1	19.64	0.46	–2.28%

	October 2011	*December 2014*	*Increase/Decrease*	*Evolution*
men –	9.7	–		
women –	10.3	–		
Bucharest	1.88	1.67	–0.21	–11.17%
men –	0.87	–		
women –	1.01	–		

§3. Political System

4. The latest Constitution, adopted in 2003, defines the Romanian Republic as a national, sovereign and independent, unitary and indivisible state. The Constitution also established Romania to be a rule of law, democratic and social state, in which human dignity, the rights and freedoms of its citizens, the free development of human personality, justice and political pluralism are guaranteed supreme values. The state is organized on the basis of separation and balance of powers – legislative, executive and judicial – as established by the norms of constitutional democracy. The legislative power is solely exercised by the Parliament, which is organized in two chambers: The Chamber of Deputies and the Senate. But the Parliament may share the legislative competence with the Government, by means of habilitation law or, in case of extreme urgency, in areas not regulated by organic laws or by the Constitution, such as: the regime of the fundamental state institution, rights, freedoms and obligations established by the Constitution, voting right or expropriations. The executive power belongs to the Government, which exercises the general leading of the public administration. The judicial power is with High Court of Cassation and Justice (hereinafter HCCJ) and the other courts established by law. HCCJ ensures unitary interpretation and application of law by all other courts, according to its competence.

5. At the moment, Romania is considered to be a semi-presidential republic. The President is directly elected by popular vote and is empowered with several rights and obligations. He ensures the representation of the Romanian State. He guards the observance of the Constitution and the good functioning of the public authorities. In this purpose he mediates between the state powers and between the state and the society. The President is entitled to nominate a Prime Minister and appoints the Government on the basis of the confidence vote given by the Parliament. He can attend the Government sessions with respect to national interest issues regarding foreign policy, defence, ensuring public order and any other cases, at the request of the Prime Minister. The President presides over the Government sessions he attends. The President has the power to dissolve the Parliament, in case the Parliament does not give its confidence vote to the Government within sixty days since the first solicitation but only after two such solicitations have been rejected. Nevertheless, this right is subject to some limitations: the Parliament can be dissolved only once in a year and cannot be dissolved during its last six months of mandate or during mobilization, war, curfew or urgency situation. With respect to foreign policy, the Romanian President signs international treaties on behalf of Romania, but such treaties must be negotiated by the Government and ratified by the Parliament. He also has attributions as regards national defence – he is the supreme commander of the army – and is able to declare, in full accordance with the law, the curfew or emergency state, but such measures need to be approved by the Parliament within maximum five days. His mandate is currently five years long but he may be suspended or impeached by the Parliament for serious deeds against the Constitution. If the Parliament approves the suspension, a referendum must be organized within thirty days in order to consult the population about his dismissal.

6. Romania is organized in forty Counties and Bucharest Municipality, which is also divided in six Districts. The local public administration is organized on principles of decentralization, local autonomy and de-concentration of public services. In those territorial-administrative units where citizens belonging to a national minority have a significant weight, provisions are made for the oral and written use of that national minority's language in the relations with the public administration authorities and the decentralized public services, under the terms stipulated by the organic laws. In cities and communes, the public authority is realized through the Local elected Councils and Mayors. At county level, there are elected County Councils which are empowered to coordinate the activity of city and commune Councils. The Government appoints for each county and the Bucharest Municipality a Prefect which shall represent the Government locally. However among the Prefects, on the one hand, the Local Councils and Mayors, as well as the County Councils and their Presidents, on the other hand, there are no subordination relationships. The powers of the Prefect in this regard are limited to the possibility to challenge in the administrative court an act of the County Council, of a Local Council or a Mayor, in case he deems they are unlawful. Such challenge from the Prefect shall de jure suspend the application of the act.

7. In Romania, justice is rendered in the name of the law, by independent, irremovable judges appointed by the President of Romania. The highest court administering justice in Romania is HCCJ. Its competences are established by the law, but its main role is to provide a unitary interpretation and implementation of the law by the other courts of law, according to its competence. It does not have any role in analysing the constitutionality of Acts passed by the Legislature.

8. Such power was granted to the Constitutional Court of Romania (hereinafter CCR) which is supposed to be the guarantor for the supremacy of the Constitution. CCR consists of nine judges, appointed for a term of office of nine years that cannot be prolonged or renewed. Three of them are appointed by the Chamber of Deputies, three by the Senate, and three by the President of Romania. Despite the fact that their status is similar to the one of a magistrate, the members of CCR must only have graduated law and have high professional competence and at least eighteen years of experience in judicial and academic activities. It is more a political than a judicial body, which has attracted severe criticism over the activity of the Court in the recent years. That is because within the extensive competences of the court one can find: the power to adjudicate on the constitutionality of laws before the enactment thereof upon notification, the power to adjudicate on the constitutionality of treaties and other international agreements upon notification, to decide on objections as to the unconstitutionality of law and ordinances, brought before courts of law or commercial arbitration, power to solve legal disputes of a constitutional nature between public authorities by request. In case the court finds that the provisions of the laws or ordinances in force are unconstitutional, such provisions shall cease their legal effects within forty-five days of the publication of the decision of the CCR. If the provisions are found to be unconstitutional before the enactment thereof, the Parliament is bound to reconsider those provisions, in order to bring them in line with the decision of the CCR. In case an international treaty is found to be unconstitutional, such treaty or international agreement shall not be ratified. The decisions of CCR are published in the Official Gazette of Romania and as of their publication, the decisions become generally binding and effective only for the future.

9. Besides HCCJ there are twelve Courts of Appeal (*Curţi de Apel*), forty-two County Courts (*Tribunale*), six Military Courts (*Tribunale Militare*) and around 300 First Instance Courts (*Judecătorii*) functioning in Romania. HCCJ is organized in four Sections – two civil, one criminal, one fiscal and administrative and the Joint Sections, having their own competences. The Courts of Appeal and the County Courts are organized in civil – criminal – labour – administrative – youth sections.

10. From 1947 until 1989 politics were dominated by the Communist Party of Romania. Only after 1989 Revolution did Romania return to political pluralism. Even so, the return was slow. In the first years of the 1990s pluralism was more formal than real, as the Front of National Salvation coagulated most of the former political elite and had a huge support from the masses, unlike the re-born 'historical' parties or the new-born parties. After the Front split up, the remaining force became first the Party of the Social Democracy in Romania and later the Social Democratic Party. This party became the most stable and powerful party on the Romanian political scene. One of the wings that split from the Front was the Democratic Party, which assumed a leftist doctrine which it changed overnight in 2004 to become a rightist Liberal Democratic Party. Such abrupt changes in the doctrine of a party or of party members and representatives should not surprise anyone as they represent one characteristic of the Romanian political organizations. The Liberal Democratic Party dominated the political scene over the past four years but had suffered a serious defeat in the latest local and national elections along with a large number of dropouts from its members. The liberal ideas were assumed by a large number of parties which finally coagulated with the historical National Liberal Party. During the past years the party has increased in support and Parliament representatives, especially after alliance with the Social Democratic Party in what became the Social Liberal Union, a coalition meant to counterweight the increasing power of the Liberal Democratic Party. The Union won the latest local and national elections in Romania (June/December 2012) and managed to form a new government. The Social Liberal Union lasted only one year due to tensions between the two leaders.

11. As a result the National Liberal Party went into opposition and in 2014 it formed an alliance with the Liberal Democratic Party under the name of Christian Liberal Alliance for the purposes of the presidential elections that took place in November 2014. The Alliance ceased to exist after winning the elections, their common candidate, Klaus Iohannis (of German origin) being now the president in office. The two forming parties are to merge under the name of National Liberal Party. The Social Democratic Party stayed in power and co-opted in the government the Democratic Union of the Hungarians in Romania. Prime Minister Victor Ponta decided to run for president, but lost the elections. As a result, the cooperation ended and the Hungarians went back in the opposition. In November 2015, after a huge fire in a nightclub, which resulted in the death of more than sixty people and sparked massive anti-governmental street movements in Bucharest, the government led by Victor Ponta resigned.[1] It was replaced by a technocrat government led by Dacian Ciolos, a former European Commissioner for Agriculture, who ran the country until the parliamentary elections set for 2016.[2] The current Prime Minister is Viorica Dancila, the first Romanian woman to hold the office, appointed by the alliance between the Social Democratic Party and the Liberal Democratic Alliance, a former branch of the Liberal Party, with support from minorities, including the Hungarians.

Local elections that took place in the summer of 2016 were mostly won by the Social Democratic Party, which also ensured, for the first time, a complete victory in the capital city, Bucharest, despite a large number of criminal cases decided or pending against prominent members of the party.[3] However, local elections also witnessed the uprising of a civic movement, the Save Bucharest Union that almost won the mayor seat in the first District of Bucharest and ended up on second place for the general mayor seat of Bucharest. The good results obtained in Bucharest led to the creation of a national movement, called Save Romania Union, which planned to rival the already established parties in the upcoming parliamentary elections.[4] However, the elections were won by the Social Democrats which scored 45% of the votes, while the National Liberal Party barely amounted to 20%. The newcomers from Save Romania Union scored almost 9%, a surprising result for a party with such a short existence, thus proving the desire of Romanians for other options.

12. Romania does not have regional parties (yet). It does have several ethnicity-based parties, despite that, according to the law, such parties are deemed to be illegal. The notable examples are the Democratic Union of the Hungarians in Romania and the Civic Hungarian Party, both of them being in a harsh competition with each other. But while the Democratic Union of the Hungarians in Romania, representing

1. http://www.bbc.com/news/world-europe-34720183, last visited on 1 Oct. 2016.
2. http://www.business-review.eu/featured/dacian-ciolos-is-romanias-new-prime-minister-91181, last visited on October 1.
3. http://www.reuters.com/article/us-romania-election-idUSKCN0YR0UI, last visited on 23 Mar. 2018.
4. http://www.intellinews.com/new-party-attempts-to-break-into-romania-s-closed-political-shop-104096/, last visited on 1 Oct. 2016.

the moderate Hungarians, managed to represent its electorate in the Romanian Parliament and to keep itself in power for twelve years (longer than any other Romanian party), the Civic Hungarian Party assumed a more radical discourse, militating openly for a territorial segregation of the Hungarian inhabited areas, and it is not represented in the Parliament.

13. Romanian nationalistic parties have not been represented in the Parliament for more than eleven years now and their electorate is definitely decreasing. But the void left by this kind of parties was rapidly exploited by populist parties, out of which, the newly emerged People's Party of Dan Diaconescu managed to be represented in Parliament after the latest 2012 elections. Last but not least, in accordance with the Constitution, all minorities existent in Romania are represented in the Parliament by one member of their ethnic group.

§4. Population and Employment Statistics

14. During the long transition from communism to democracy and from planned economy to market economy, the Romanian labour market had suffered significant changes with respect to the volume and the structure of main indicators of labour force. Such changes included the decrease of the active and employed population, an approximate constant level of unemployment rate, but also an exodus of workers – mainly blue collars – towards the western states of the European Union (hereinafter EU). Unemployment rate increased severely after the financial crisis of 2008 struck Romania.

15. Starting with 2002, the active population oscillated around 10 million people. In 2011, it reached 9.6 million out of which 96.1% belonged to a range of employees between 15 and 64 years old. On the other side, if during 2005–2008 the employed population had continuously increased, from 2009, the employed population started to decrease, reaching in 2011 the lowest value ever recorded: 9.1 million. Out of these 55% were male. Also, until 2002 the majority of the employed population was represented by the population living in the rural areas. After 2003, this majority was replaced by the population domiciled in the urban areas, namely 55.59% in 2011. The unemployed reached 0.73 million people in 2011, although this figure is most likely lower than the real one. Even so, it is a 7.3% increase from 2009 and the most worrying fact is that out of these almost one-third (28.8) are youngsters aged 15 to 24 years. As for the second trimester of 2014, the unemployment rate reached 6.7% maintaining the trend of high unemployment rates for those within the 1,524 years range: 23.9%.

16. The active population of Romania is in steady decline, due to low natality and migration. Based on the latest available figures, published by the National Institute of Statistics for the fourth quarter of 2017, the active population of Romania counted 8.96 million people, out of which 8.54 million people were employed and 425 thousand were unemployed. In January 2018, the unemployment rate at national level was 3.99, the smallest in the last twenty-five years, although the figure covers only those registered as unemployed with the National Labor Agency.[5]

5. http://www.anofm.ro/files/Comunicat%20de%20pres%C4%83-%20Rat%C4%83%20%C5%9Fomaj%20Ianuarie%202019.pdf, last visited on 23 Mar. 2018.

Unemployment Rate for 4th Quarter of 2017 Based on Age, Gender and Residential Area

20.1/22.7	5.9/4.0	3.3/2.4	3.7/3.3	2.9/2.0	27.5/17.7	4.7/5.7	2.3/3.7	3.1/4.1	3.1/2.0
15–24 years	25–34 years	35–44 years	45–54 years	over 55 years	15–24 years	25–34 years	35–44 years	45–54 years	over 55 years
Male/Female					Urban/Rural				

Source: According to the data made available by the National Institute for Statistics of Romania.[6]

6. Press Release concerning the unemployment rate for the 4th Quarter of 2017 for population aged between 15–64 years), available online, in Romanian at: http://www.insse.ro/cms/sites/default/files/com_presa/com_pdf/somaj_tr4r_17_0.pdf, last visited 23 Mar. 2018.

17. In 2016, out of the total number of employed persons, 23.1% worked in the agricultural sector, 29.9% in the industry or construction sectors and 47% in the service sector. Compared to 2015, the trend showed a decrease of people employed in the agricultural sector, as well as in education, public administration and defence, with an increase in manufacturing, construction, trade and hotel/restaurants sectors.[7]

18. Romania is a country of small businesses. Despite the large proportion of employees acting in the private sector, the majority of the employers are small enterprises with less than ten employees.

7. https://www.cotidianul.ro/creste-rata-de-ocupare-in-munca-a-populatiei-in-varsta/, last visited on 23 Mar. 2018.

Active Economic and Social Operators of National Economy, by Size Class

Economic and social operators type	Of which: by size class, according to employees number				
	Total	0–9	10–49	50–249	250 and over
2012					
Total	850,049	776,525	58,519	12,781	1,680
Enterprises	472,342	413,803	48,275	8,584	1,680
– Agricultural holdings	16,680	13,658	2,170	228	24
– Enterprises from industry, construction, trade and other services	449,482	394,024	45,601	8,251	1,606
– Financial and insurance companies	6,780	6,121	504	105	50
Public administration	13,135	542	8,194	3,882	517
Private administration	86,494	84,102	2,050	315	27
Private entrepreneurs	278,078	278,078	–	–	–
2014					
Total	910,317	838,513	57,399	12,223	2,182
Enterprises	507,582	449,512	47,943	8,416	1,711
– Agricultural holdings	17,471	14,928	2,301	217	25
– Enterprises from industry, construction, trade and other services	483,066	428,191	45,142	8,092	1,641
– Financial and insurance companies	7,045	6,393	500	107	45
Public administration	12,847	709	8,192	3,501	445
Private administration	90,984	89,388	1,264	306	26
Private entrepreneurs	298,904	298,904	–	–	–

Active Economic and Social Operators of National Economy, by Size Class

Economic and social operators type	Of which: by size class, according to employees number				
	Total	0–9	10–49	50–249	250 and over
2012					
2015					
Total	921,133	848,155	58,107	12,622	2,249
Enterprises	513,989	455,561	48,053	8,624	1,751
– Agricultural holdings	18,396	15,768	2,375	229	24
– Enterprises from industry, construction, trade and other services	488,210	433,075	45,177	8,277	1,681
– Financial and insurance companies	7,383	6,718	501	118	46
Public administration	12,935	349	8,486	3,634	466
Private administration	97,061	95,097	1,568	364	32
Private entrepreneurs	297,148	297,148	–	–	–

§5. Social and Cultural Values

19. Romania, as it is today, is a new-born state. Its unification process started with a personal union of two principalities – Walachia and Moldavia – in 1859, and was completed sixty years later, at the Peace Treaty of Paris, in 1919 as a result of the First World War. The borders were again changed in 1940 when Romania was forced under the pressure of Nazi Germany to give up territories to Hungary, Bulgaria and the USSR and again in 1947, after the Peace Treaty of Paris, when solely the western border with Hungary was restored, while the southern and eastern borders remained as they were in 1940.

20. The unification process coincided also with a movement for modernization. After being under the rule or domination of the Ottoman Empire, the Romanian principalities turned their attention towards the western models, especially towards France. A Civil Code, based on the French model, was adopted, later a Commercial Code based on the Italian counterpart. Foreign investments were attracted and a process of industrialization started especially after huge reserves of oil were discovered. Although such process was slow and the economy was still more agricultural based, the economic increase of Romania was steady.

21. As part of the modernization process, Romania chose to become a laic state, nationalized the huge estates of the church and established private property for the peasants working those estates under the rule of Prince Alexandru Ioan Cuza (1858–1866). Such process was continued after the First World War when the large domains were also abolished (1921). The initiative, despite its good intentions, was not really that successful in economic terms, as the properties were too small to enable economic development, but it ensured in most cases the possibility to self-support. It also developed a powerful sense of property within the society.

22. When the unification process started, the Romanian principalities were ruled by a prince of national origin. Taking Belgium as a model, and with extensive support from France, Romania tried to bring a prince from a foreign family. It was believed that such an initiative would ensure the support of a foreign power and also would provide the heavily divided Romanian society with a neutral arbiter. Thus, a German prince of family Hohenzollern was elected to rule Romania in 1866. Under his and his family's rule, Romania gained its independence (1877), became a constitutional monarchy (1881) and continued – with successful results – the abovementioned unification process (1912, 1918).

23. The emergent state, monarchy and economy were all brought to a harsh end after 1947 when Romania became part of the communist world. Monarchy was abolished, so were the multi-party system, private property, market economy and other basic freedoms. The society suffered an important transformation as economy became planned, private initiative was discouraged, everything started being dictated from the centre and a rule of terror was imposed. The country was subjected to large-scale industrialization, despite lacking rough materials and resources. On the positive side, the infrastructure was developed and the level of alphabetization increased.

24. In 1989, Romania experienced the only violent revolution against communism in Europe. The communist leader was executed on Christmas and it seemed that the country would return to its former values and path. Despite such beliefs and hopes, monarchy was not restored, no lustration was ever performed, the nationalized and confiscated properties were returned to their former owners solely after ten years from the Revolution and the transition was longer and tougher than expected. The society proved not to be mature enough to handle freedom and market economy. Late privatization, corruption and lack of professional knowledge caused Romania to remain one of the poorest and underdeveloped economies in Europe. Solely the will to rejoin the civilized world and to restore the western values remained alive and culminated with the country's joining the European Union in 2007. Also the efforts made to fulfil the application process and the access to the main freedoms of the Union slowly transformed and shaped Romania into a more democratic state.

25. Without the minimum historical background provided above, it would be impossible to understand the Romanian reality and mentality as it has suffered too many, radical changes, in a relatively short period of time.

26. Romania currently defines itself as a democratic and social state, governed by the rule of law, in which human dignity, the citizens' rights and freedoms, the free development of human personality, justice and political pluralism represent supreme values. According to the constitution, the social and economic foundations are the *right of association* – into political parties, trade unions, employer's associations and other forms of association – *labour and social protection of labour* – everyone has a free choice of his/her profession, trade or occupation, as well as work place, and is also entitled to measures of social protection (such as employees' safety and health, working conditions for women and young people, setting up of a minimum gross salary per economy, weekends, paid rest leave, equal wages for equal work), right to strike and the prohibition of forced labour – *right of private property* – which is guaranteed and protected by the law for both national and foreign citizens, prohibition of expropriation for other reasons than public utility and in the absence of prior and just compensation, prohibition of nationalization – last but not least, *economic freedom* – which establishes free access of persons to an economic activity, free enterprise and the guarantee of these rights' exercised under the law.

27. By defining itself as a social state and by granting large rights and powers to trade unions, combined with the communist influence, the last decade of the twentieth century was affected by a large number of strikes, resistance to privatization attempts and also state interventionism in the economy. This has changed after 2003, when the ties with the EU became stronger and Romania prepared for accession. Even more, after the financial crisis in 2008, Romania has undertaken some of the hardest economic measures, including the salary cuts of 25% in the public sectors, increase of taxes, including Value Added Tax (VAT) from 16% up to 24%, and has also amended the protectionist Labour Code to provide employers with more power over their employees in order to make the labour market more flexible, collective bargained agreements were abolished, etc. with almost no response from trade unions. With such harsh measures implemented by the state itself many doubt that Romania is still a social state.

28. It can be said now that Romania is a pluralistic society in which all citizens and assemblies are allowed the freedom to promote their own interests and enjoy a large number of legal safeguards. Though it is not a perfect society and still has to deal with a lot of issues, the progress made is undeniable. Unlike other western states and unlike the 1990s, conflict and strife seems to have disappeared from Romania, but it is impossible to say that there is no more collision of interests and that harmony has settled over. It is simply the sign of a change in the perception of all constituencies, while the weakening of the trade unions just follows the European trend.

Chapter 2. The Economic System

29. Based on the constitutional provisions and the current realities, the economic system of Romania may be best described as a social market system: a free market economy, based on free enterprise and competition, in which the state undertakes to secure: free trade, protection of fair competition, protection of the national interests in economic, financial and currency activity, stimulation of the national scientific and technological research, arts and protection of copyright, exploitation of natural resources, protection of environment, creation of all necessary conditions so as to increase quality of life and also implementation of regional development policies in compliance with the objectives of the European Union.

Chapter 3. The Legal System

§1. Legal Family or Tradition

30. The Republic of Romania is now a parliamentary democracy, though, as stated before it is more of a semi-presidential republic. According to the Constitution now in force (2003), the Parliament is the supreme representative body of the Romanian people and the sole legislative authority of the country, while the president has the role of representing the Romanian State and of safeguarding the national independence, unity and territorial integrity of the country. He is also deemed to act as a mediator between the Powers in the state as well as between the state and society.

31. The state is organized on the basis of the now classic principle of separation and balance of powers: legislative, executive and judicial, within the framework of constitutional democracy. Romania is in fact still a centralized state, local autonomy expressly excluding any kind of reference to territorial autonomy, despite the requests of the Hungarian minority. Until now, the unitary and indivisible character of the Romanian state, as expressed in the Constitution, remains untouchable.

32. However, public administration in territorial-administrative units is claimed to be based on the principles of decentralization, local autonomy and de-concentration of public services. Such autonomy is, as in any other unitary state, purely administrative and not political, as in federal or regional states. In subsidiary there is also a financial autonomy, but in a very limited sense. To sum up, for the moment, the local autonomy in Romania is functional and not territorial, thus there are no local governments, no local parliaments or local courts in Romania. Also, there is no acceptance of segregation on ethnical criteria.

33. In order to accommodate the Hungarian minority, the Romanian legal system has established that in the territorial-administrative units where citizens belonging to a national minority have a significant weight they shall be entitled to use their national minority's language in the relationship with the local public authorities and the decentralized public services. The Romanian citizens belonging to a national minority are also entitled to express themselves in their mother tongue before the courts of law, though such right is restricted to oral and not-written expression. Last but not least, Romania established a large cultural autonomy, which enables the members of national minorities to learn in their mother tongue, such right being expressly recognized by the Constitution.

34. Romania has established the prevalence of the Universal Declaration of Human Rights and any treaties and international conventions with respect to citizen's rights and liberties to which it is party, by stating that in case there are inconsistencies between the constitutional provisions and such treaties or conventions, the latter shall take precedence, unless the former or the national laws comprise more favourable conditions.

35. Romania implemented the former Civil Code in 1865, following the Napoleonic Code and the Italian Civil Code. It was amended many times throughout its existence, but it was the only code unaffected by the communist modifications. Its provisions with respect to freedom to contract, general principles with respect to obligations, etc. have remained intact and have governed the economic relations in Romania. It was finally replaced by the New Civil Code (NCC) in 2011, inspired by the Civil Code of Quebec, considered one of the most modern civil codes in the world and also suitable for the Romanian realities. The Commercial Code was implemented in 1887, following the model of the Italian Commercial Code from 1882 but including provisions from the German and Belgian legislation as well. It was applied solely before and in between the two world wars. Starting with 1947 the communists applied it only for acts of commerce involving an element of extraneity. After 1989 it was rediscovered and re-applied in the domain to which it was destined. It was completed by the Company Law No. 31/1990, by Law No. 64/1995 regarding the judicial reorganization and bankruptcy, and later by Law Nos 85/2006 and 85/2014 regarding the insolvency procedure.

36. The court system comprises only national courts. HCCJ has the authority to issue Guiding Decisions (*Decizii de îndrumare*) which become mandatory country-wide for all courts from the moment of their publication in the Official Gazette of Romania. In accordance to its legal competences it can also annul or modify decisions of the Courts of Appeal. The latter are also entitled to annul or modify the decisions of County Courts, when they do not act as first instance courts, while County Courts have the power to annul or modify the decisions of the First Instance Courts. The constitutionality control lies with the CCR, an institution established on political basis but whose decisions may influence not only the legislative but also the judiciary. The effects of both the Guiding Decisions of HCCJ and the Decisions of CCR apply only for the future.

37. As in the case of any other EU Member State, Romania began to be affected more and more by community law. Prior to its accession, in order to comply with all the requirements meant to facilitate the integration process, Romania has amended its laws with respect to securities, competition law, consumer protection, taxation, European common policies with respect to agriculture, commerce and transportation. In the energy field, company law, insurance, intellectual property, banking, Romania has also implemented the European directives. Following the increased specialization and familiarization of Romanians with the European legislation applicable at union level, Romania has faced the reality of having internal rules and legislation stricken down by the Court of Justice of the European Union for being contrary to the basic freedoms of the EC Treaty such as free movement of goods, persons, services and capital.

§2. Relation and Hierarchy Between the Sources of Law

38. From a purely legal standpoint, the sources of law are normative acts, legal customs and the case law.

39. Among the sources of law, statutes occupy the highest place in their hierarchy.[8] The term 'statute' has two meanings: a broad and a narrow one. Broadly, the term 'statute' expresses any of the forms of normative acts – a decision, a decree, an order or a directive – that was issued by a state body and has a mandatory character. The most adequate expression for all forms or categories of sources of law is normative (legal) act.[9] In a narrow sense, the term 'statute' signifies the strongest source of law, the normative act that is superior to all other normative acts – the one adopted by the supreme legislative body, the Parliament.[10]

40. A statute is defined as an imperative rule which expresses a norm. In a wider sense, it is that imperative rule, of general character, that originates from an authority, it is imposed on a person from the exterior and it is sanctioned through public force.[11]

41. In the category of 'statutes' as superior sources of law, three forms are possible:

(1) The Constitution and/or fundamental statutes.
(2) Statutes and Codes.
(3) Normative acts with legal power: decrees, government ordinances, certain international acts (treaties, conventions, agreements that were ratified by the Parliament).

42. The Romanian legislator classified statutes by their hierarchy and power,[12] and divided them as follows: (a) Constitutional or fundamental statutes – in Romania this would be the Constitution of 2003; (b) Organic statutes and (c) Ordinary statutes.

43. The Constitution's supremacy is understood in the sense that all state bodies, including the Parliament, the Head of State or the Government, are held to abide the Constitution and that its modification is subjected to special procedures established by the Constitution itself,[13] qualified majority and, sometimes, a referendum. The principles of constitution's supremacy and constitutionality of laws and other

8. A terminological clarification is needed. Romanian language uses for 'law' in the general sense (a system of rules recognized in a particular country) or as a field of study (civil law, criminal law) the term drept, while for 'statute' as a normative act, it uses the term lege. However, lege is often translated as 'law,' which in this case would have been confusing. Hence, for the purposes of this chapter, the monograph opted to use 'statute' as normative act and 'law' as system or area of study, although in other chapters law is used to define a normative act as well.
9. Mihaela-Cristina Paul et al., Elemente de teoria dreptului pentru economisti (General Theory of Law Elements for Economists) § 1, 91 (1st ed., Ed. Hamangiu 2014).
10. *Ibid.*, at 91.
11. *Ibid.*, at 91.
12. Article 73 of the Constitution.
13. Articles 150–152 of the Constitution.

normative acts stem from the principles of state power supremacy and state sovereignty, represented by the state's legislative body, the Parliament, as an expression of the organization and functioning of a rule of law state.[14]

44. Organic statutes are an intermediary category between constitutional and ordinary statutes. This is a special category due to its object that concerns either the organization, functioning and structure of different state bodies or special areas of social life. They are adopted with the votes of majority of members of each Parliament chamber, unlike ordinary laws that are adopted with the majority of the present members of each Parliament chamber.[15] Organic statutes cover areas such as: the electoral system; the organization, functioning and financing of political parties; organization of referendums; the state of siege and emergency; criminal offenses, penalties and execution thereof; statute of public servants; organization of local public administration; statute of national minorities and, of specific importance to this monograph, the general status of property and inheritance.

45. Ordinary statutes are all other statutes elaborated by the Parliament, following the regular enactment procedure, and covering various domains and social relations that do not fall in the area of the Constitution or of organic statutes.

46. In the category of statutes as sources of law one includes also Codes of laws, which are in fact a set of rules, systematized in a certain form and vested with the power of law by which a larger area is regulated. Normally, they contain ground rules of a certain field of law, for which reason they are also designated as 'general law' as opposed to a special statute, of recent nature, designated as 'special law.'[16]

47. Customs are rules of conduct that appear spontaneously, due to their repeated application for a relatively long period of time in a human community. State power refers to customs in various manners: recognition and sanctioning (thus empowering those deemed useful and needed to consolidate the rule of law), acceptance and tolerance (of those that do not require transformation in a normative act and that do not contradict the established legal order) or banning (of those that contravene the established order). Only recognized and sanctioned customs represent a valid source of law, though currently its importance is greatly reduced and limited to international law (maritime law, for example).[17]

48. The case law is more and more present among the sources of law in Romania, mainly due to recent reforms of civil and civil procedure codes, an aspect which is perceived as a shift towards the common law systems.[18] Guiding decisions of the HCCJ together with the rulings of the CCR are mandatory for all courts and have similar effect to a statute. The same effect occurs with the decisions of the European

---

14. Paul, et al., 2014, p. 93.
15. Article 76 of the Constitution.
16. Paul, et al., 2014, p. 94.
17. *Ibid.*, at 88.
18. *Ibid.*, at 97.

Court of Human Rights and those of the European Court of Justice that are binding on national courts. It should be mentioned here that given the recent and modern nature of the NCC, most of the previous holdings of the Guiding Decisions of the HCCJ and the CCR rulings were included already in the text of the NCC. Since 2011, when the NCC came into force, and until the writing of this monograph, there are only three Guiding Decisions regarding the interpretation of the NCC (the two of them relevant to this monograph have been mentioned) and no ruling of the CCR regarding the unconstitutionality of the NCC.

49. The architecture of the new codes also seems to point out a shift towards a case law based legal system: the New Code of Civil Procedure (hereinafter NCCP) has turned the mechanism for preliminary questioning into an integral part of Romanian procedural law.

50. However, the mandatory effect is limited only to the aforementioned types of decisions. The remaining case law, although important, regards only concrete cases and cannot issue general and permanent solutions. From this perspective, the formal role of case law among the sources of law is limited to a minimum,[19] despite efforts to create a unitary practice among courts.

19. *Ibid.*, at 98.

Part I. Regulatory Framework Concerning Electricity

Chapter 1. General Introduction

51. The law that provides the general framework applicable to the electricity field in Romania is Title I (Electricity), Chapters 1–9 of Law No. 123/10.07.2012 concerning Electricity and Natural Gas.[20] Title I of the aforementioned law basically brings into the national law the provisions of the Directive No. 2009/72/CE issued by the European Parliament and Council on 13 July 2009, with regard to the common regulations for the internal market of electricity and the annulment of Directive 2003/54/CE, published in the Official Journal of European Union series L no. 211/14.08.2009, the provisions of Directive 2005/89/CE of European Parliament and Council from 18 January 2006 with regard to the actions aimed to guarantee the safety of electricity supply and investments in infrastructure, published in the Official Journal of the European Union series L no. 52 / 21.02.2004. The aforementioned provisions became enforceable on 3 March 2013.[21]

52. According to Article 96, paragraph 2, the Law No. 123/2012 explicitly annuls the provisions of the Electricity Law No. 13/2007, published in the Official Gazette of Romania No. 51/23.01.2007, with the exception of the provisions of Articles 7–11 which contain certain data related to the functioning and the duties of the competent authority,[22] National Regulatory Authority for Energy (NRAE),[23] in

20. Law No. 123/2012 concerning electricity and natural gas (hereinafter Electricity Law) was published in the Official Gazette of Romania, Part I, No. 665/21.09.2012. It was rectified in 2012 and subsequently amended in 2013, 2014, 2015 and 2016.
21. *See* the provisions in Art. 96. para. 1 of Electricity Law.
22. According to the provisions of Art. 3, point 2, with regard to the meaning of certain terms and expressions used in Title I from Electricity Law, competent authority refers to the National Regulatory Agency for Energy (hereinafter referred to as NRAE).
23. Article 7, para. 3, provides that NRAE has the possibility to establish its own territorial structures with no legal personality, Art. 8, para. 2 with regard to financing the activity of the competent authority stipulates NRAE's right to establish the level of the tariffs and contributions for a period of minimum four years, having the possibility to update them with the inflation index. The Article also mentioned NRAE's obligation to make the established tariffs and contributions public. Article 9 stipulated the way in which the competent authority is managed and the appointment procedure. Article 10 included provisions with regard to hiring and the salaries of NRAE's staff. At the same time, Art. 10, para. 3 explicitly established NRAE's staff obligation to maintain confidential the information obtained while performing or as a result of their work attributions. In the end, Art. 11 details NRAE's attributions, which will be dealt with in the section regarding NRAE.

the field. With regard to the related documents,[24] the law stipulates that until NRAE establishes new regulations according to the provisions of the new law, all the normative acts issued on the basis of the previous law will remain valid, under the condition that they do not contravene the provisions of Electricity Law.

53. According to Article 97, paragraph 3, the relevant ministry,[25] will perform, in twenty-four months since the new law enters into force, an evaluation of the operation of the transmission and system operator based on the model 'system-independent operator' and, if it deems necessary, it will be able to make a proposition to the Government for adopting the model of separation of property, according to Article 9, paragraph 1 of the Directive 2009/72/CE. The basis of the separation of property consists of the following four conditions required by the European law:

(1) all undertakings which own transmission systems act as transmission and system operators;
(2) the same person(s) is (are) not allowed to exercise direct or indirect control over a company that performs any of the activities of production or supply, to have direct or indirect control or to exercise any right over a transmission and system operator or over a transmission system, and on the other hand to exercise, directly or indirectly, any rights over a company that performs any of the activities of production or supply;
(3) it is not allowed that the same person/s appoint members in the surveillance committee, administration committee or in other bodies that legally represent the company in case of a transmission and system operator or in case of a transmission system, and likewise, to exercise, directly or indirectly, the control or any other rights over an undertaking that performs any of the activities of production or supply;
(4) it is not allowed that the same person be member in the surveillance council, administration council or in any other body that legally represents the company, in the case of companies that perform any of the activities of production or supply, as well as in the case of a transmission and system operator or a transmission system.

54. The declared purpose of the above-mentioned four conditions is, according to the stated Recitals Nos 4, 9, 15 and 18 of the Directive 2009-72/CE: to create discrimination-free access to the network, which cannot be achieved without the effective separation (unbundling) of the networks from the activity of production and supply.

24. Article 96, para. 5 of Electricity Law.
25. According to the provision of Art. 3, point 36, with regard to the meaning of certain terms and expressions used by Title I of Electricity Law, the relevant ministry refers to the Ministry of Economy, Commerce and Business Environment.

55. According to the Directive 2009-72CE, the purpose of the separation of property is to grant complete independence of the exploitation activities of the networks from the interests related to production and supply as well as to prevent the exchange of confidential information. Nevertheless, the separation of the property is to be achieved by protecting the interests of the shareholders, who have the option of separating either by direct assignment, or by dividing the integrated shares in shares of the company integrated in the two separated companies.

56. The Directive 2009/72/CE offers the EU Countries the possibility to opt for *complete* separation of the property on their territory, but Romania has not implemented this model yet.

§1. Principles Governing Electricity Sector

57. The Electricity Law establishes the framework for the activities performed in the field of electricity and thermal energy produced in cogeneration with the purpose of optimizing the primary resources. The latter one must be performed by simultaneously following the principles below:

- accessibility;
- disponibility;
- supportability;
- following the safety, quality and environmental protection regulations.

58. The Electricity Law does not apply, on the one hand, to electric chargers, portable electricity generator sets, electric installations placed on any types of vehicles and, on the other hand, to stationary sources of electricity, on direcţ current, unless they are connected by power invertors to the National Energy System (hereinafter referred to as 'NES').

§2. Fundamental Purposes

59. All the activities in the electricity field and the thermal energy one produced in cogeneration must follow a series of *fundamental purposes* explicitly stipulated by law: ensuring the long-term development of the national economy; diversifying the basis of the primary energetic resources; creating, ensuring and improving the functioning of the competitive market of electricity; ensuring, according to the regulations, the non-discriminatory access to all the participants in the electricity market and in the public electricity networks; price and tax transparency; making safety stocks for the necessary combustibles; ensuring the interconnected functioning of NES with the electro-power system of the neighbouring countries; promoting the use of new and renewable resources of energy; protection of the environment; preventing the sabotage and terrorism against NES infrastructure.[26]

§3. The Energy Strategy and Policy: The Energy Programme

60. The Romanian law chose to incorporate in the law dispositions with regard to the energy strategy and policy as well as with regard to the energy programme. The said dispositions mainly restate the fundamental principles and objectives related to the performance of the activities in the energy field, which are dealt with as a purpose in itself.

61. With regard to the *national strategy*, it is stated that it shall define the objectives of the electricity field for medium and long term, as well as the modalities of their implementation, under the conditions of ensuring a long-term development of the national economy. The national strategy must be issued by the relevant ministry and must be approved by the Government which has to consult in advance the non-governmental organizations, social partners and business sector representatives. The text of the law only stipulates the obligation *to consult* the related parties therefore, as a result, it can be inferred that the ministry in charge does not have the obligation to consider their opinions.

62. In 2017, Romania updated its energy strategy, after almost a decade, by revising its development projections of the energy sector on both long and medium term, by answering to the latest challenges and technological developments and by reconsidering the role of the state. The general objective of the strategy in the energy field is to ensure the necessary energy for the present time as well as for a medium and long term, at the lowest price achievable, adapted to a modern market economy and a civilized life standard, under quality and safe supply conditions and based on the long-term development principles. Romania's energy strategy for the next fifteen years focuses on increasing the quality of energy services, which should reflect in the living standard and the increase of economic competitivity. At the same time, the new strategy is consumer oriented, seeking to address the consumer's future need for quality products and services, for energy generation that is less polluting, while maintaining optimal parameters in energy supply.[27] Another main objective is to achieve a mix of generation sources that would guarantee Romania's energy security. The latter is now defined as a balance between energy independence (granted by the efficient exploitation of national resources) and interconnectivity with neighbouring countries and regional energy markets, as an alternative, including in situations of crisis.[28]

26. The complete list of the basic objectives is detailed in Art. 2 of Electricity Law.
27. *See* Romania's Energy Strategy for 2016–2030 with a view towards year 2050, page viii, available on line at: http://www.mmediu.ro/app/webroot/uploads/files/2017-03-02_Strategia-Energetica-a-Romaniei-2016-2030.pdf, last visited 23 Mar. 2018.
28. *Ibid.*

63. In a similar way, the *energy policy* is developed by the relevant ministry for a medium period of time and must follow the directions established by the national strategy[29] as well as the ones of the government's programme, considering the probable evolutions on long term and with the consultation in advance of the related parties. Thus, the energy policy must consider the following objectives:

– increasing the level of energy security;
– competitive energy markets, as the basis of a competitive economy;
– clean energy, with low emissions and decreased pollution levels;[30]
– modernizing the system of energetic governance;
– protecting the vulnerable consumer and reducing energy poverty;[31]
– prioritizing the electricity supply for the isolated inhabited places;
– developing the international energy-related cooperation;
– participating in regional markets and in the European market of electricity.

64. For the application and implementation of the energy-related strategy and policy, the Electricity Law stipulates that the finance sources will be ensured by the undertakings in the field from: their own funds,[32] from the state budget, from the local budget, from loans or grants. Regarding the financial resources considered for the implementation of the proposed measures, the National Strategy mentions, among others: resources obtained from the privatization and the restructuring of companies in the electricity field; financial resources obtained by the companies in the energy field from their own sources or from bank loans obtained either from external financing bodies, or banks, without the guarantee of the Romanian state; financial resources obtained by incorporating companies with mixed capital; financial resources obtained by using specific mechanism promoted by the Kyoto Protocol in addition to the environment policy; and financial resources obtained from structural funds awarded for financing projects in the energy field.[33]

65. However, with regard to support – such as state aids for the implementation of the energy strategy and policy – it is mentioned that they will be granted only

29. *Ibid.*, p. 12.
30. *Ibid.*, p.13.
31. Idem, p. 14. In addition, *see* The National Plan of Action for Power Efficiency (PNAEE) available online at: http://www.minind.ro/energie/PNAER_final.pdf and The National Plan of Action in Renewable Power Sources Field (PNAER) available online at: http://www.minind.ro/energie/PNAEE_var_finala.pdf, both last visited 23 Mar. 2018.
32. According to the provisions in Art. 3, point 42, with regard to certain terms and expression used in Title I of Electricity Law an undertaking in the electricity field refers to any natural or legal person who carries out at least one of the following activities: electricity production, transportation, distribution, supply or purchase and has commercial, technical and/or maintenance attributions related to these activities. Final customers are explicitly excluded from the above-mentioned definition.
33. *See* Romania's Energy Strategy for the period 2016–2030 with a view towards the year 2050, pp. 7–8 and 18–19, available online at: http://www.mmediu.ro/app/webroot/uploads/files/2017-03-02_Strategia-Energetica-a-Romaniei-2016-2030.pdf, last visited on 23 Mar. 2018.

under the provisions of the law. The national strategy includes, in turn, measures for the reduction of the social impact, especially with respect to vulnerable consumers.[34]

66. The energy policy represents a programme which includes stimulation measures for the investment activities, research and development, long-term development, efficient use of energy resources and of energy efficacy as well as other activities meant to ensure the safe and secure functioning of NES. The programme is approved by Government Decision and is enforced by the Government, the relevant ministry and other specialized bodies of the central and local public administration.

34. *Ibid.*, pp. 14, 21 and 50–54.

Chapter 2. Authorities and Competences

§1. The Relevant Ministry: Duties

67. As stated already, the relevant Ministry in the electricity field is the Ministry of Economy, Commerce and Business Environment.[35] It issues the national energy strategy, the energy policy and ensures its implementation.

68. Electricity Law presents the main attributions[36] of the relevant ministry, which include:

- establishing the institutional frame necessary for companies in order to be able to perform their activities in the electricity field;
- issuing programmes and measures plans for the application and implementation of the governmental policy in the electricity field;
- representing the state, as contracting authority, in concession agreements in the electricity field;
- supervising the implementation and the compliance with the established measures for the environment protection;
- defining the critical infrastructure for NES;
- promoting and facilitating, together with NRAE, the cooperation between the operators on the electricity market, the transmission and system operators and their correspondents at regional level, including the cross-border aspects, with the purpose of creating a competitive market for the electricity;
- drawing up, every second year, starting with October 2010, a report concerning the embodiment of the established national targets and the measures implemented to facilitate the access of electricity produced from renewable energy sources to the electricity network;[37]
- elaborating and notifying to the European Commission the National Action Plan in the Area of Renewable Energy.[38]

§2. Regulatory Authority

69. The regulatory authority in the electricity field is NRAE, organized and functioning as an autonomous administrative authority, with legal personality.

35. *See infra* fn. 40.
36. *See* Art. 6 of Electricity Law.
37. Article 20, para. 4 of Law No. 220/2008 for setting up the support scheme of electricity production from renewable energy sources, republished in the Official Gazette of Romania, Part I, No. 577/13.08.2010 and subsequently amended in 2011 and 2013 (hereinafter Law No. 220/2008).
38. Article 21, para. 1 of Law No. 220/2008.

According to the provisions of the Electricity Law NRAE's organization, functioning, competence and attributions are established in conformity with the law, namely the Law No. 13/2007 of Electricity, whose provisions related to NRAE are still in force.[39]

70. NRAE drafts, establishes and follows the implementation of the mandatory regulations at national level, necessary for the functioning of the field and of the electricity market, under conditions of efficiency, competition, transparency and consumer protection. To this end, NRAE has the following duties and powers:[40]

- to establish mandatory regulations for the undertakings that activate in the electricity field;
- to grant, modify, suspend or withdraw authorizations and licenses for the undertakings that activate in the electricity field;
- to create and approve the calculation method necessary for establishing the prices according to the regulations;
- to establish the prices applicable to the captive consumers;
- to approve the prices between the undertakings of the electricity field on the regulated electricity market, prices for system services, transmission and distribution of electricity, prices for activities and services related to the production of thermal energy in cogeneration intended for the population, with the purpose of ensuring the protection of the final client;
- to establish the framework contracts for supply, the ones between undertakings with the purpose of sale, purchase, transmission, system service, electricity distribution, as well as the ones with the purpose of selling thermal energy produced in cogeneration;
- to issue the regulations for electricity supply to the consumers and obtain its approval by the Government;
- to approve the technical and commercial regulations for the undertakings in the field;
- to control if the undertakings in the electricity field comply with the issued regulations, with the price system in force and to apply sanctions in case of infringement, as well as to establish a methodology for following and controlling the above. The general frame related to the investigation procedure, the rights of NRAE's staff and the method for obtaining judicial authorization are provided by Articles 84–90 of Electricity Law and are detailed in the Guidelines regarding the organization and the performance of the investigation, approved by NRAE president order and completed with the applicable dispositions of the Criminal Procedure Code;[41]
- to establish the procedure for solving the pre-contractual disagreements and to solve the disagreements derived from concluded contracts between undertakings in the electricity field, for solving the contracts of electricity supply and the contracts for connection to the network;

39. *See* Art. 96, para. 2 of Electricity Law.
40. For a complete list of NRAE's attributions, *see* Art. 11, para. 2 of Law No. 13/2007 of electricity.
41. *See* the provisions of Art. 90 corroborated with those of Art. 88, para. 6 of Electricity Law.

– to notify the relevant ministry and the Competition Council with regard to the abuse of a dominant position on the market and with regard to the infringement of the legal provisions related to competition, whenever they deem that the regulations for competition and transparency have been infringed;
– to monitor the electricity market with the purpose of evaluating its efficiency, transparency and competition level and the continuity in supply, based on their own established regulations and to submit annual reports to the prime minister and the relevant ministry regarding the identified problems and the implemented solution.

71. The monitoring activity of NRAE in the electricity field mainly refers to: regulations for interconnection facility management and allocation, congestion management in NES, the duration for the performance of the connection to the transmission and distribution network by the operator and the duration for reconnection after repair, the terms, conditions and prices for the connection of new producers of electricity, in order to guarantee that they are objective, transparent and non-discriminatory, the way in which the transmission and system operators as well as the distribution operators undertake their obligations as provided by law, the transparency and competition level of the electricity market operation. It also monitors the development and functionality of the Green Certificates market and draws up, annually, a report concerning the functioning of the support scheme for electricity produced from renewable energy sources.[42]

72. In order to fulfil its aforementioned duties, NRAE collaborates with the Competition Council, with the National Authority for Customer Protection, with the ministries and other related special bodies from central or local public administration, as well as with the associations of energy consumers and undertaking in the field. In addition, NRAE has the right to access information and documents related to the activity field of the undertakings, as well as their accounting records.

73. With respect to the support scheme for production of electricity from renewable energy sources, NRAE has the following competences:

– It accredits groups or power plants which benefit from the Green Certificates support scheme, on the basis of the Regulation for accreditation of electricity producers from renewable energy sources.
– It notifies the decision of accreditation to the applicant and to the operators involved.
– It elaborates the regulatory framework for monitoring costs and production revenues of electricity producers from renewable energy resources which benefit from the Green Certificates support scheme.[43]

42. Article 20, para. 1 of Law No. 220/2008.
43. Article 6, para. 6 of Law No. 220/2008.

– It sets up and updates periodically an electronic *Register of power plants accredited for obtaining Green Certificates.*[44]

44. Article 24, paras 1 and 2 of the Regulation for accreditation of electricity producers from renewable energy sources for the application of the Green Certificates support scheme (hereinafter Regulation for accreditation), published in the Official Gazette of Romania, Part I, No. 477/28.06.2014.

Chapter 3. Authorizations and Licenses

§1. Activities Requiring Authorization and License

74. The Romanian law separates the activities that require authorization from the activities that require license. Thus, the creation of new power facilities or the reengineering of the currently existing ones is to be carried out based on an establishment authorization, granted under the conditions of the law, while the activities of production, transmission, system service, distribution and supply of electricity, as well as the activities of administration of the electricity market are to be carried out based on license.[45]

75. With regard to the creation of new energy facilities of electricity production, including facilities for the production of electricity and co-generated thermal power or their upgrading, it is necessary to obtain an authorization for establishment only if the electricity installed of the respective installed capacities is over 1 MW[46] under the condition that those facilities be installed on/in buildings that are public or private property of the state or of territorial and administrative divisions, on the private property of the requester or on other properties owned by any legal title.[47] If the electricity installed is between 500 kW inclusively and 1 MW inclusively, it is only necessary to inform the competent authority with regard to the investment project and to periodically submit a report on the status of the execution. If the electricity installed is below 500 kW, the obligation to notify and report lies with the operators of electricity distribution.

76. The activities that require to be licensed by NRAE are: the commercial exploitation of the facilities that produce electricity (including cogeneration), the electricity transmission service, the system service, the distribution service, the administration of electricity market and the supply of electricity. The license for the supply activity will be necessary as well for commercial activities with the electricity obtained based on a processing agreement. According to the law, only one license is to be granted for the electricity market operator and one for the balancing market operator.[48] Similarly to the establishment authorization for new facilities, the licenses for commercial exploitation of new power facilities will be granted only if they are located on state property or are owned as property or with any other legal title.

77. The Electricity Law also provides for exceptions from the licensing regime. Thus, the owner of production facilities of electricity which can be started without NES tension, self-starting groups and which are used for the purpose of safety supply with electricity of their own equipment or installations can be commercially

45. *See* Art. 8, paras 1 and 2 of Electricity Law.
46. *See* Art. 10, para. 1, letter a) of Electricity Law.
47. *See* Art. 10, para. 3 of Electricity Law.
48. *See* Art. 10, para. 2, letter f) of Electricity Law.

exploited without license. In addition, the household client, owner of electricity production facilities, respectively electrical and thermal in cogeneration, connected to the network, with a power lower than 100 kW can commercially exploit them without license. The distribution service can be performed without a license either by the owner of one/more electrical distribution network/s, if from the technical connection permit results that the approved power for the users of this/these network/s adds in total below 1 MW, or by the owner of one/more electrical distribution network/s, if its/their users hold consumption points exclusively inside of one/more buildings supplied by that/those network/s.[49] Moreover, the activity of a natural or legal person of reselling the electricity to the users of the electricity network under exploitation is to be carried out without license. Any other activity or service carried out without license or establishment authorization under the conditions of the law represents a misdemeanour and is punished with a fine.[50]

§2. Authorization and Licensing Procedures

78. The procedure of granting, modification, suspension and withdrawal of the authorizations and licenses, the terms and conditions of granting consisting of criteria, authority limits, certifications, approvals, guarantees etc., differentiated according to categories of facilities and activities under authorization are provided for in the regulations issued and approved by order of NRAE's president.

79. According to the law, the authorization procedure is started by the applicant's submission of an application to the competent authority. The request will enclose the documents which certify that the applicant meets the economic, financial, technical and professional conditions established for each of the categories of power facilities and activities in the field. As a preventive measure, the law explicitly excludes from right of authorization the applicants that have as controlling shareholders or as administrator persons who were previously controlling shareholders or administrators of licensed companies that failed to pay the obligations resulted from transactions on the electricity market.

80. In establishing the conditions for granting the authorizations and licenses for new production facilities, the following criteria are considered: safety and security of the electrical systems, of installations and related equipment; protection of public health and safety; environmental protection; field occupancy; location choice; public domain use; energy efficacy; nature of primary sources; technical, economic and financial characteristics of the applicant; conformity with consumer's protection measures; contribution to the formation of facilities to achieve the global European objective according to which the power from renewable sources is to represent 20%

49. *See* Art. 10, para. 5, letters d) and e) of Electricity Law.
50. *See* Art. 93, para. 1, point 1 and Art. 93, para. 2, point 1 letter b) as well as point 2 letter b) of Electricity Law.

of the final gross power consumption of EU in 2020; contribution to creating power production facilities for the reduction of toxic emissions; efficient and rational value of the primary energy resources.

§3. Suspension and Withdrawal of Authorizations and Licenses

81. The failure of the establishment authorization or license holder to meet their legal obligations, or to obey the conditions, limitations, restrictions, interdictions or the tasks established by the authorization, respectively by the license, can be established by the competent authority ex officio, upon the notice of third parties or upon notification by a holder, in which case the procedure will be differentiated depending on the infringement being or not imputable to the holder and depending on it being reversible or not. Thus:

- if it is not imputable to the holder, the competent authority will decide either to grant *a term of compliance*, in case the situation created is remediable, or to *withdraw the authorization, respectively the license*, in case the situation is not remediable;
- if it is imputable to the holder, the competent authority will decide either *the suspension of the authorization, respectively of the license*, for a determined period of time during which the holder can rectify the situation and become conformant with the authorization, respectively the license, in case the situation is remediable, or the *withdrawal of the authorization, respectively of the license*, in case the situation is not remediable.

82. Without regard to the previously described situations, NRAE will withdraw the establishment authorization, respectively license, if the holder is disqualified, incapacitated or bankrupt, if the concession or the rental of the power facility is terminated, as well as if the holder has sold the power facility.

§4. The Challenging Procedure

83. The refusal to grant an authorization, the lack of response within the legal term or any other solution considered illegal or damaging, can be challenged in court. The jurisdiction belongs to the first instance of the Bucharest Court of Appeal, Administrative Litigation Division, and the procedure is the one established by the Law No. 554/2004 on administrative litigation[51], corroborated with the dispositions of the Code of Civil Procedure in force. Even though the Electricity Law mentions only the applicant as the owner of the challenging petition, in my opinion such a

51. Law No. 554/2004 concerning administrative litigation was published in the Official Gazette of Romania No. 1154/7.12.2004 (with all subsequent amendments).

delimitation is debatable, having in consideration that the activities subjected to authorization and licensing are of public interest[52] and thus, they can be challenged by any interested party.

§5. The Rights and Obligations Arising from Authorizations and Licenses

84. The Electricity Law imposed limitations on the property right over the lands and goods which are either public or private property, of natural or legal persons, and on the activities performed by natural or legal persons in the vicinity of these electrical facility, in favour of authorizations and licenses holders[53] who have the following rights:

- the right of use of the land for performing the works necessary for the implementation of the object of the authorization;
- the right of use of the land for ensuring the normal functioning of the facility representing the object of the authorization;
- an easement right of underground, surface or air passage for the installation/dismantling of electricity networks or other related equipment for the power facility and for access to the place of installation;
- the right to obtain the limitation or cancellation of some activities that might endanger persons and assets;
- the right of access to public utilities.

At the same time, they have the right of use or a legal easement right over the lands that are in the property of third parties, which are included in the safety and protection areas.[54]

85. The object of the rights of use and trespassing is the public utility, thus these rights are automatic, carried out free of charge and without registration to the Land Register during the entire existence of the power facility or, temporarily, in the event of upgrading of a running facility, the repair, revision or intervention activities in case of breakdown. At the same time, the holders of the use and trespassing rights over the public or private property of the state or administrative and territorial units will be exonerated from the obligation to pay taxes, duties or any other obligations instituted by the central and local government authorities.[55] Furthermore, the holders of authorizations and licenses can perform activities of vegetation clearance or

52. *See* Art. 12, para. 1 of Electricity Law which stipulates: 'The establishment or upgrading power facilities for which authorizations are granted, as well as the activities and the services for which licenses are granted, are of public interest, with the exception of those which are exclusively aimed to the authorization or license holder's own consumption.'
53. *See* Art. 12, para. 2 of Electricity Law.
54. *See* Art. 15, para. 3 of Electricity Law. The same article, first paragraph, defines the protection areas and the safety areas as those areas established: 'in order to protect the normal functioning of the power facility and the enclosures, as well as to avoid jeopardizing persons, goods and environment'.
55. *See* Art. 12, para. 13 of Electricity Law.

modelling cuts for creating and maintaining the minimum requesting distance from the electricity networks, by specialized staff, obeying the legal provisions in force.[56]

86. The exercise of the same rights over private properties will be done based on a frame-agreement[57] and with the payment of compensations and indemnities[58] whose content and calculation method will be included in a methodology approved by the Government. Under the provisions of Energy Law the owners of the lands that constitute the object of the rights of use or easement can request to the license and authorization holders to conclude a frame-agreement, and license and authorization holders have the legal obligation to conclude such an agreement within maximum thirty days from the request date.

87. The amount of the compensation will be established by the agreement of the parties, or, in case of disagreement, by the court. It is important to note that the payment of the indemnities and compensation resulting from the limitations to the property rights do not affect in any way the rights to compensation of the property owners for any damage caused by the holders of authorizations and licenses while carrying out their right of use or easement.

§6. Obligations for Holders of Authorizations and Licenses

88. The holders of authorizations must, on one hand, apply protective measures for the persons, assets and environment during the entire duration of works, and on the other hand obtain all the permits, agreements and certificates prescribed by law in order to execute the object under authorization.

89. The license holders must, during the entire validity period of the license, comply with the conditions set in the granted license; must keep separate accounting data for each activity that makes the object of the license;[59] must constitute and maintain financial guarantees to allow them to perform the activity and to ensure the continuity of the service; must provide the competent authorities with the necessary information for the proper performance of its activities; and to keep, to submit for audit and make public the annual accounting records at company level (according to the disposition of the Fourth Directive 78/660/CE of the Council from 25 July 1978). With regard to the obligation of undergoing an audit, this will verify

56. *See* Art. 12, para. 12 of Electricity Law.
57. According to the provisions of Art. 12, para. 5 of Electricity Law, the frame convention includes procedure rules with regard to the terms and conditions regarding the period, the content and the limitations to the rights of usufruct and trespassing.
58. According to the provisions of Art. 12, para. 10 of Electricity Law, when establishing the indemnity the following criteria will be considered: the land surface, the types of crops and plantations (as well as the additional improvements affected by the works) and the activities limited due to the works.
59. If the undertaking in the power field is horizontally or vertically integrated, they must proceed as if the activity is performed by different undertakings and to make financial reports in the format requested by the competent authority. *See* Art. 13, para. 2, letter b) of Electricity Law.

whether the obligation of non-discrimination is complied with or not and the cross subsidies between the activities carried out by the audited undertaking.[60]

§7. Rights and Obligations of Authorization and License Holders over the Property of Third Parties

90. The use right over the land for the execution of the activities necessary for the execution, relocation, dismantling or upgrading of the energy facility is valid during the entire period necessary for the execution of works. In exercising this right, the holder of the authorization, can, under the conditions of the law: store on the land materials, equipment, machines and installations, which are necessary for the executions of works, to remove or reduce crops, plantations, constructions or any other existing facilities, remove materials, capture water, install and use machines, install offices and construction site houses,[61] stop or reduce the owner's activities, to the extent of what is strictly necessary for the executions of works for the authorized facility, with a written notice to the affected owner or service provider.

91. The use right for the above-mentioned purpose terminates either when the duration for the execution of works expires, or previous to this moment, if the works are finished with anticipation, are suspended or the authorization is waived. In any of these cases, the owner has the obligation to immediately notify the owner.

92. With regard to the *use right* for the land *for ensuring the normal functioning of the energy facility*, it is valid throughout the entire period when the facility is running and the holder is entitled to exercise it whenever necessary for ensuring a normal functioning of the facility. Thus, the holder has the right: to store materials, equipment, machines, equipment for necessary maintenance, revisions, repairs and interventions; to install and use machines; and to remove or reduce crops, plantations, or any other existing facilities and to reduce the owner's activities, under the condition of issuing a previous written notice in the owner's attention.[62]

93. The trespassing right for underground, surface or air includes the right of access and the right of execution of the works on the site where the power facility is located with the purpose of upgrading, repair, revision or technical checks. Moreover, in order to avoid endangering the safety of goods or services performed in the execution area, the holder of the authorization or of the license can obtain the limitation or the suspension, during the entire execution period, of the activities performed in the nearby area by other persons, with prior written notice containing the start and the end date of the works.

---

60. *See* Art. 13, para. 4 of Energy Law.
61. In order to set-up equipment, offices and temporary lodging, the holder must obtain the prior agreement of the owner. In my opinion this right can be explicitly awarded through the frame-agreement or by a separate convention.
62. With the exception of the breakdown cases, when the owners are to be notified as soon as possible.

94. The holder of the authorization or of the license who has access right to *public utilities* must exercise it in good faith and in a reasonable way, without causing any prejudice to others' right to access the public utilities. In addition, the owners of authorizations and licenses have the common obligation to clear the land and restore it to its initial state.

§8. The Right to Request Expropriation

95. As previously mentioned, the necessary land for the establishment and the functioning of the power facility can be the private property of a third party or of the holder of the authorization, or it can be public property. If the land is the property of a third party, the applicant for the authorization of establishment has as first legal option, the possibility to buy the land from the owner or to initiate the legal procedure of expropriation of the land with the purpose of public utility, by compensating the owner under the provisions of the law, and thus to obtain its concession for the period during which the power facility exists.[63]

96. Needless to say, the expropriator will be the state which, as a result of the expropriation, will become the owner of the land and will be able to grant concession to the requester under the conditions of the law. The advantages of having the right to request the expropriation lay in the fact that, in case the negotiations for purchasing the land fail, by expropriation, the applicant can immediately start the performance of works, as their activity is not affected by the state setting and paying compensations to the expropriated party.

§9. Access to Electrical Network: Exception

97. The license holders and final clients have regulated access to the public electrical networks, which represent a mandatory service that must be carried out by the transmission and system operator, as well as by the distribution operator.

98. As an exception, the access to the network can be restricted if the connection influences NES's safety or if the transmission and distribution operator does not have the necessary facilities. In either case, the refusal of access must be motivated and justified, based on objective criteria, from both technical and economic point of view, according to NRAE's Regulations of 2 August 2013.[64] Resolving any conflict with regard to the restriction of the access to the network lies within NRAE's competence which will issue a mandatory decision for parties within two months from the receipt of the claim.[65]

---

63. Also *see* the subsection regarding the Concession.
64. The Regulation concerning the connection of users to electrical networks of public interest of 2 Aug. 2013 was published in the Official Gazette of Romania, Part I, No. 517 bis/19.08.2013.
65. *See* Art. 25, paras 1–3 of Electricity Law.

Chapter 4. Production

§1. Procedures for New Power Plants: Public Tender

99. In case the production facilities built on the basis of the authorization procedure are insufficient to guarantee the safety of the internal consumption supply, the relevant ministry can initiate a public call for tenders or any other similar procedure with the purpose of awarding the contract, through which to invite new undertakings or pre-existent production license holders to make offers for establishing new power plants. The winner is awarded an establishment authorization and a license.

100. The procedure for the tender as well as the list containing the criteria based on which the offers will be chosen and the award granted are to be approved by Government Decision, at the proposal of the relevant ministry. The responsibility for the organization, monitoring and control of the offer procedure lies with the transmission and system operator.

101. The law establishes the following principles[66] that must be considered for the issue of the tender procedure: the information with regard to the procedure must be published in the Official Journal of the European Union within at least six months prior to the final date for submitting the offers; each interested undertaking must be offered access to the tender book; the tender book must contain a detailed description of the specifications of the contract and the procedure; the complete list of the selection and awarding criteria, including aids or subventions; and to take in consideration the offers of electricity supply with long-term guarantees, proposed by the existing production units, in the case of the calls for the tender procedure which refer to the production capacity predicted to be necessary.

§2. The Concession Agreement: Object and Granting Authority

102. According to the provisions of the Electricity Law the object of energy concessions can be: assets which are public or private property of the state; the public activities; and services of national interest from the electricity sector. The granting authority for the assets which are in the public or private property of the state or for the public activities or services of national interest will be the relevant ministry,[67] which means that for the assets which are in public or private property of the local public authorities or for the public activities and services of local interest, the granting authority will be the local authority.

66. *See* Art. 18, para. 5 of Electricity Law.
67. *See* Art. 19, para. 2 of Electricity Law.

103. The general legal framework for the concession agreements, the granting procedure as well as the standard content for the tender book are issued by the granting authority according to the provisions of the law and are approved by Government Decision. With regard to the granting conditions, maintaining, suspending and withdrawal of the concession, they are established by the concession agreement. The concession agreement enters into force on the date when it is published in the Official Gazette of Romania, Part I.[68]

104. The production of electricity and thermal energy in cogeneration is done by undertakings that own electricity production units. The producers may benefit of guaranteed free access or of priority access to electrical networks according to NRAE's regulations.

§3. The Obligations of the Producers

105. The Romanian law stipulates a series of main obligations for the producers of electricity, however the list is not complete. Among these obligations one can find: ensuring the supply of electricity and technological system services (following the conditions requested by licenses, agreement clauses and regulations in force); public and non-discriminatory offers on the competitive market of the entire available electricity; submitting to the transmission and system operator a physical notice in case of negative imbalance against the concluded agreements, with the exception of the producers who benefit of a support programme; maintaining a fuel reserve at a sufficient level for covering the production and continuous supply of electricity; being in compliance with the requests of the transmission and system operator; as well as submitting an annual activity report to the competent authority.

§4. The Rights of the Producers

106. The phrasing of Article 27 from the law stipulates that electricity producers *may* benefit of guaranteed and priority access to the electrical network and implies that this access is merely a right of the producers and not an obligation for the owners of the electrical networks (as it appears the latter have the right to deny access to the former). However, Article 30 clears the ambiguous phrasing of Article 27 and stipulates that the main right of the producers is the access to electrical network, therefore my interpretation is that, actually, the access cannot be denied discretionarily.

107. In addition, the producers have the right to obtain a passage corridor for their own electricity lines; to make transactions with the electricity and the technological system services on the regulated and competitive market, in a transparent and non-discriminatory manner; to create and maintain their own telecommunication system; to trade the thermal energy produced in cogeneration; and to access

68. *See* Art. 19, paras 3–5 of Electricity Law.

support programmes available at national level. Moreover, they can also quote technological system services on the European internal market, as long as they do not endanger the safe functioning of NES. Similarly to the list for the obligations, the list of rights is not exhaustive, either.

§5. Electricity Produced from Renewable Energy Sources and in Highly Efficient Cogeneration

108. By its geographical location, Romania benefits from a high and diversified potential of renewable resources, equally distributed on the country's territory. Nevertheless, these resources will be exploited on a larger scale only once the performance-price ratio of technologies will improve. In the past six years, Romania has advanced the usage of an important part of its wind and solar power potential, although the investment rhythm is expected to slow down in the near future.[69] Romania has achieved the EU target regarding renewable energy for 2020[70]. However, this target was reached due to a pro-investor support scheme, which meant a considerable cost for end-consumer.[71] Hence, the drafters of the Romanian Energy Strategy do not envision the introduction of new scheme for support of renewable energy production, before 2030, when the one in place will expire.[72] The expectation is that in the next decade, wind and solar power related technologies will become competitive even in the absence of a support scheme. Therefore, the drafters expect investments to continue and wind and solar power generating capacities to increase.[73]

109. The law defines as renewable power sources: wind power, solar power, waves and tide power, geothermal power, hydroelectricity, power contained in the biodegradable fraction of products, waste and residues from agriculture, forestry and related industries (biomass), power contained in the fermenting waste gas (waste dump gas), power contained in the fermenting sediment gas in the used waters treatment installations, power contained in secondary gas products obtained through fermentation of organic residual materials (biogas), power contained in liquid products obtained from the distillation of fermented organic material (alcohol fuel) and power obtained from other renewable sources, not currently exploited.[74]

110. Among these, the most convenient renewable resources, regarding costs, resources volume and technologies necessary for the electricity production are the hydroelectricity plants (including micro hydroelectricity plants), wind turbines and

69. Romania's Energy Strategy 2016-2030, with a view towards year 2050, p. 26, available online at: http://www.mmediu.ro/app/webroot/uploads/files/2017-03-02_Strategia-Energetica-a-Romaniei-2016-2030.pdf, last visited on 23 Mar. 2018.
70. *Ibid.*, p. 107.
71. *Ibid.*, p. 108.
72. *Ibid.*, p. 108.
73. *Ibid.*, p.108.
74. *See* Art. 67 of Energy Law.

plants in cogeneration that use biomass, although the strategy identifies significant investment needs to replace old power plants[75]

111. Besides private capital, important financing sources are those made available through European investment programs, both structural and strategic investment funds, or those provided by investment and development banks. At the same time, an important role is to be played by public-private partnerships, regarding investment schemes such as ESCO[76], for increasing the energy efficiency of buildings.[77] Finally yet importantly, a possible source of financing is the budgetary income associated with tenders for emission permits related to the ETS[78] system.[79] The access to funds for promoting electricity from renewable sources and in highly efficient cogeneration is done by following the approval and certification rules issued by the competent authority,[80] after their approval by the European Commission, under the provisions of the law.[81]

112. Although the Romanian law prohibits the cross-subsidy between an undertaking's regulated activities as well as the non-regulated ones, in case of electricity production in highly efficient cogeneration or from renewable power sources which benefit of funds at national level, the cross-subsidy will be analysed between the production activity in cogeneration or from renewable power sources which benefit of the funds and the other activities carried out by the undertaking.[82]

113. NRAE establishes by technical and commercial guidelines, on the one hand, the guaranteed access to the electrical networks and priority dispatching of electricity produced through renewable power sources and in highly efficient cogeneration, and on the other hand, the priority access to electrical networks and priority dispatching of electricity produced from renewable power sources and in highly efficient cogeneration in plants with installed power lower or equal to 1 MW, to the extent in which the safety level of NES is not affected.

114. The expenses for the promoting criteria of the electricity produced from renewable power sources and in highly efficient cogeneration, take in consideration the following: the reach of national target regarding the percentage of the electricity produced from renewable power sources, the compatibility with the competitive principles on the market, the characteristics of different renewable power sources and the electricity production technologies of the electricity, as well as the promoting of efficient usage of renewable sources.

75. *See* Romania's Energy Strategy 2016–2030, with a view to 2050, p. 93.
76. ESCO as defined by Romania's Energy Strategy 2016–2030, with a view to 2050, means Energy Services Company, *see ibid.*, p. vi.
77. *Ibid.*, p. 93.
78. ETS as defined by Romania's Energy Strategy 2016–2030, with a view to 2050, means Emission Trading System, *see ibid.*, p. vi.
79. *Ibid*, p. 94.
80. *See* Art. 72, para. 2 of Electricity Law.
81. *See* Art. 74 of Electricity Law.
82. *See* Art. 79, para. 5 of Electricity Law.

115. The electricity produced from renewable sources and in highly efficient cogeneration is certified by origin guarantees issued by NRAE.

116. The regulated prices for the electricity and thermal energy produced in highly efficient cogeneration plants or from renewable sources that benefit from funds are the objects of the provisions of Government Decision[83] regarding the approved support programmes.

§6. Electricity Produced from Renewable Energy Sources. General Framework

117. In 2008, the Parliament adopted Law No. 220/2008 for setting up the support scheme for electricity production from renewable energy sources[84] which implemented a support scheme for investors in renewable energy sources which would help Romania reach the targets agreed with the EU. The purposes were: to increase the security of electricity supply and to reduce the imports of primary energy sources by integrating into the national energy balance the renewable energy sources, to support and stimulate the creation of new workplaces, local and regional development; to reduce pollution; to ensure the co-financing for attracting external financing sources for promotion of renewable energy sources; to define the norms concerning guarantees of origin, the applicable administrative procedures and connectivity to the electricity network for the electricity produced from renewable sources; and to establish the durability criteria for biofuels and bio-liquids.

118. The Romanian legislator chose a support scheme through Green Certificates which benefited from a rigorous regulation coming from the mentioned law but also from numerous Decisions, Approvals, Orders and Guidelines issued by the competent authorities. The scheme promoted the use of electricity produced from renewable sources by reducing the production costs, by either increasing its sale price or by implementing purchase quantities of this type of energy[85] and by implementing a mandatory quota system[86] combined with the trade of Green Certificates.

119. Basically, the Romanian state offers, on the one hand, a formal evidence attesting that the electricity was really produced and, on the other hand, allows these certificates to be traded at a minimum set price, as commercial goods, on a specialized market, as it will be later shown.

---

83. *See* Art. 79, para. 7 letter a) of Electricity Law.
84. *See supra* fn. 18.
85. Article 4, para. 4–5 of Law No. 220/2008 corroborated with the provisions of the Guidelines to establish the mandatory annual quotas of electricity produced from renewable energy sources which benefit from the Green Certificates support scheme and of Green Certificates acquisitions, published in the Official Gazette of Romania, Part I, No. 909/15.12.2014, derogated by the provisions of Art. III, para. 1 of Law No. 23/2014 for the approval of Government Emergency Ordinance No. 57/2013 concerning the modification and amendment of Law No. 220/2008, published in the Official Gazette of Romania, Part I, No. 184/14.03.2014.
86. Article 4, paras 5, 7, 8 and 9 of Law No. 220/2008.

I. Beneficiaries of the Support Scheme

120. The support scheme is applicable to electricity delivered into the electricity network and/or to customers if it is produced from renewable energy sources as defined by Electricity Law and Law No. 220/2008.[87] However, the duration of the support scheme varies depending on the type of electricity production.[88] Additionally, in order to benefit from the support scheme, electricity producers from renewable energy sources need to be accredited in accordance with the provisions of the law[89] and if they obtained the origin certificates for their fuels or raw materials, issued by the competent Ministries.[90] In order to be accredited, power plants need to fully comply with the legal requirements set by Law No. 220/2008, Regulation for accreditation, but also with those set by the Electricity Metering Code.[91]

121. Once the NRAE approves the accreditation,[92] it notifies its decision by fax or email, in a scanned version, to the accredited undertaking and to the following entities: the network, distribution, transmission and system operators to which the accredited power plants are connected; the transmission and system operator – Transelectrica, for the issuance of Green Certificates;[93] and the Green Certificates Market operator – OPCOM – for registering on the market. In addition, NRAE posts and updates on its website the list of accredited undertakings and power plants for the purposes of the support scheme.[94]

II. National Target Levels Concerning the Percentage of Electricity Produced from Renewable Energy Sources

122. The national target levels concerning the share of electricity produced from renewable energy sources in the final gross consumption[95] for 2015 and 2020 is 35% and, respectively, 38%. Romania reached its target level for 2020 in 2014,

87. *See* Art. 67 of Electricity Law corroborated with Art. 3, para. 1 and para. 6 of Law No. 220/2008.
88. Article 3, para. 2 and para. 8 of Law No. 220/2008.
89. Article 3, para. 3 corroborated with Art. 6, para. 6 of Law No. 220/2008 corroborated with Art. 10 of the Regulation for accreditation.
90. Article 3, paras 3, 4, 9 and 10 of Law No. 220/2008. The Procedure for Issuance of Origin Certificates for biomass originating from forestry and ancillary industries was published in the Official Gazette of Romania, Part I, No. 360/28.05.2012, and was amended in 2013.
91. Electricity Metering Code is available online, in Romanian only, at http://www.anre.ro/download.php?f=f657gg%3D%3D&t=vdeyut7dlcecrLbbvbY%3D, last visited on 23 Mar. 2018.
92. Article 4, para. 6 of Law No. 220/2008.
93. According to Art. 6, para. 1, the transmission and system operator issues monthly green certificates to the producers of electricity produced in power plants which use renewable energy sources, from which one deducts the electricity quantity required for ensuring their own technological consumption.
94. Article 9 of the Regulation for accreditation.
95. The national objective concerning the share of electricity from renewable energy sources in the final gross consumption is 24% for 2020. By national gross consumption from renewable sources is calculated according to Art. 5, para. 2 of Law No. 220/2008 corroborated with the provisions of the NRAE Guidelines.

much sooner than it was anticipated. Although the support scheme has proven successful from this perspective, it was heavily criticized by the industry for raising the manufacturing costs and thus making the Romanian products more expensive and less competitive on the internal market.[96] The reason behind both the success and the criticism was that the price for Green Certificates was really high, and once the target was reached, their price would have to maintain itself at its maximum level. In other words, consumers had to pay more for their electricity bills, which would have caused a serious social and economic crisis.[97]

123. The Romanian Government thus intervened and amended the law. With respect to the mandatory quotas benefitting from the support scheme, the law provides that for 2015–2020, the annual mandatory quota of electricity will be established annually through Government Decision, at the proposal of the Energy Department, within sixty days from the moment when it is notified by National Agency for Mineral Resources (NAMR). For the period 2020–2030, the quotas are set by the relevant ministry and approved by Government Decision, but they cannot be lower than the level of 2020.[98]

124. NRAE establishes by Order, each December, the annual mandatory quota of Green Certificates that needs to be purchased, estimated for the following year, the number of Green Certificates estimated to be issued on the basis of the information concerning the estimated electricity to be produced from renewable energy sources and also the final electricity consumption estimated for the following year.[99] By March, NRAE establishes the mandatory quota of Green Certificates acquisition corresponding to the previous year, on the basis of the quantities of electricity produced from renewable energy resources, but without exceeding the quota set through Government Decision.[100]

NRAE establishes within forty-five days from the end of each trimester, for the previous trimester and for each economic operator who is obliged to purchase Green Certificates, on the basis of the number of purchased Green Certificates and the electricity invoiced to final customers or self-supplied, the level of fulfilment of the mandatory annual quota of Green Certificates for the previous trimester.[101]

---

96. *See* Catalin Stanescu, *Romania: The Target for Renewable Energy Production for 2020 Has Already Been Reached – But Is It Really Good News?* 6 Renewable Energy L. Policy Rev., 93 (2015).
97. Idem at 93.
98. Article 4, paras 4^2 and 5 of Law No. 220/2008.
99. Article 4, para 7, as modified by Law No. 122/2015 concerning the approval of certain measures concerning the promotion of electricity from renewable energy sources as well as the amendment and completion of certain normative acts, published in the Official Gazette of Romania, Part I, No 387/03.06.2015 (hereinafter Law No. 122/2015).
100. Article 4, paras 4, 7 and 9 of Law No. 220/2008.
101. Article 12, para 1^1 of Law No. 220/2008, as modified by Law No. 122/2015.

Chapter 5. Transmission of Electricity

125. The transmission of the electricity is done by the transmission and system operator, organized as authorized legal person by NRAE, under the provisions of the law in force. The electricity transmission network of electricity is the public property of the state, together with the fields on which it is located[102] at the time when the Electricity Law came into force, with the exception of those fields and electricity transmission network components for which the transmission and system operator gained property rights under the conditions of the law.[103]

§1. The Certification of the Transmission and System Operator

126. The transmission and system operator is organized by the model 'independent system operator' and has that status of a juridical person certified by NRAE under the legal conditions.[104] Namely, the conditions for certification are those included in Articles 9 and 10 from the law. In its turn, the certification procedure is completed according to Article 3 from the Regulation (CE) No. 714/2009, and the appointment of the transmission and system operator as independent system operator is approved by the European Commission, as a result of the communication between NRAE and the certified undertaking. NRAE also has the obligation to notify the European Commission with regard to any application for certification of a transmission network owner or of a transmission and system operator who is controlled by a person or by several persons from one or more third party countries, as well as any circumstances that could lead to one or more persons from one or more third party countries gaining control over a transmission network or over a transmission and system operator, if the case.

§2. The Certification of the Transmission and System Operator Controlled by Third Party Countries

127. If the transmission and system operator is controlled by third party countries, NRAE will request the prior approval of the European Commission with regard to that company complying with the certification conditions and that their certification would not endanger the safety of power supply for the European Union. If the response of the European Commission is favourable, NRAE will have a final decision within two months from the issue of the notice. According to Article 32, paragraph 2, from Energy Law, the lack of a response from the European Commission will be considered a tacit approval. It is important to mention that the approval of the European Commission is only a recommendation, and NRAE is not obligated to follow it, being able to issue an opposing decision, with proper motivation.

102. According to Art. 30, para. 30, second thesis of Electricity Law, the land is to remain public state property during the entire existence of the network.
103. *See* Art. 30, paras 1–4 of Electricity Law.
104. The conditions for certification are included in Art. 34 of Electricity Law.

128. However, from the phrasing of the law, it results that NRAE cannot refuse the certification at its own discretion, but only on the conditions explicitly provided for by Article 32, paragraph 4, namely: the entity in discussion does not fulfil the certification conditions, and the granting of the certification may place in danger the safety of electricity supply within the national or the European Union territory. For deciding upon this latter cause for refusal, NRAE must take into consideration: the rights and obligations of the European Union towards that country and other specific elements of the case and the third party country that is involved.

§3. The Re-evaluation of Certification of the Transmission and System Operator

129. As any change in circumstances can lead to losing the conformity of the certification with the legal requirements, the transmission and system operator is under the obligation to notify NRAE of any planned transaction as well as of any situation based on which one or more persons from one or more third party countries could gain control over the transmission system or over the transmission and system operator.

130. In addition to the above-mentioned situation, the re-evaluation of the certification conditions can be decided by NRAE, ex officio or at the motivated request of the European Union.

§4. The Obligation of the Transmission and System Operator

131. The transmission and system operator has the obligation to issue investment and development plans of the transmission network for a period of ten years, according to the actual status and the future evolution of the power consumption and of the sources, including the import and exports of power. The investment plans must include financing and implementation methods of the investment in the transmission networks, taking into consideration the management plans and systemization of the land through which they pass, under the environment protection norms approved by NRAE.[105]

132. The obligations and main activities carried out by the transmission and system operator are set forth in Article 36 from the Electricity Law. Thus, the transmission and system operator provides a public service of transmission for all users of the electricity transmission network, without discrimination, ensuring access to any applicant under the conditions of the law.

133. While normally, the transmission and system operator is forbidden to sell electricity, the law established the exceptions in which such transactions would be allowed, namely: for covering the consumption in their own electrical networks and

105. *See* Art. 35, paras 1–3 of Electricity Law.

consumption sites in their possession, as well as for ensuring the balance of production–consumption by sales/purchase transactions on the balancing market, or by sales/purchase transactions with other transmission operators from neighbouring countries. However, the transmission and system operator can carry out sales/purchase transactions of technological system services without limitation, under the regulations issued by NRAE. The electricity transactions, as well as the technological system service transactions must be performed based on transparent and non-discriminatory procedures, by competitive mechanisms.[106] The same principle of non-discrimination must be applied when setting the prices for transmission and distribution of electricity produced from renewable energy sources and the one produced from conventional energy sources.[107]

134. With regard to the activities performed by the transmission and system operator, the law mentions only the most important ones, such as: ensuring the long-term capacity of the transmission network for answering the reasonable request of transmission and electricity; exploiting, maintaining, upgrading and developing the transmission network in economic conditions, in order to ensure its safety, reliability and efficiency, with full compliance of environmental protection rules; contributing to the safe supply of electricity by ensuring a suitable transmission capacity; ensuring the management of energy flux within the transmission network; purchasing the technological system services; qualifying the dispatchable producers and customers who provide technological system services, based on their own procedure, approved by NRAE; exchanging information with other interconnected transmission and system operators; ensuring the capacities for interconnection; exploiting, maintaining and developing installations for measuring the electricity transfer through the transmission networks and at the interface with the users of the electricity transmission networks which they own; analysing and certifying if the users of the electrical transmission network comply with the technical conditions, according to the provisions of the technical guidelines in force.

135. At the same time, the transmission and system operators provide the transmission of the power measurement results to the electricity market operator in order to reimburse the transactions from the balancing market and the imbalances of the parties responsible for the balancing. It also provides access of transmission service beneficiaries in order to verify the measurement sets. They have to carry out the operational planning and operative management of NES, at central and regional level, based on their own forecast, attaching priority to the production installations that use renewable energy sources and that produce highly efficient electricity in cogeneration, to the extent permitted by the safe operation of the electricity national network.

106. *See* Art. 36, paras 1–4 of Electricity Law.
107. Article 9, para. 4 of Law No. 220/2008.

136. With respect to electricity produced from renewable energy sources, the production facilities will be connected to the electricity transmission and distribution network, provided the safety of NES is not affected. The law provides that electricity produced from renewable energy sources has priority access to the transmission and distribution network,[108] a measure meant to encourage investments in renewable energy power plants. Network operators must make available for each new producer of electricity from renewable energy sources, who wishes to connect to the system, a set of complete information, including: a detailed estimation of connection costs; a reasonable and precise deadline for receiving and analysing the connection request; an estimative calendar for each proposed connection to the electricity network and information concerning the financial guarantee for the connection.[109] The transmission and market operator must warrant the transmission of such energy, by ensuring the reliability and safety of the electricity networks.[110]

137. In addition, the transmission and system operator authorizes the staff that performs the operative management, collects, records and archives statistical data with regard to NES's operation; drafts and submits for approval to the competent authority the technical guidelines and specific regulations necessary for the operative management activity, after consultation with the participants on the electricity market; issues the plan for NES's protection against major perturbations; endorses studies, programmes and works regarding NES's development; issues and submits for NRAE's approval the regulations for congestion management, including the interconnection facilities, as well as the norms for granting the interconnection facilities; and organizes and controls the balancing market for electricity.

138. In case of emergency situations or major breakdown that would jeopardize the safety of NES operation, the transmission and system operator may dispose for a limited period of the water reserve outside the water storage exploitation programmes, under the obligation to notify the competent water management authorities.

139. In order to ensure an efficient competition and an efficient operation of the electricity market, the transmission and system operator as well as the transmission network owner will make public the information regarding their *own* activity, but they are forbidden to disclose sensitive commercial information obtained during their activities, including information from third parties gathered while granting or negotiating their access to the network.

140. The transmission and system operator is obliged by law to submit a report, once every three months, with regard to the physical flow of electricity that took

108. Article 25, paras 1 and 1^1 of Law No. 220/2008.
109. Article 25, para. 2 of Law No. 220/2008.
110. Article 9, para. 1 of Law No. 220/2008.

place as electricity import from third party countries from the European Union, during the previous three months before the date of the report. This report is to be submitted to NRAE and to the relevant ministry in order to inform the European Committee.[111]

§5. Duties of the Owner of a Transmission Network

141. According to the law, the owner of a transmission network has the obligation to cooperate with the transmission and system operator, supplying them with all the relevant information necessary for the latter to be able to fulfil all their above-mentioned obligations. In addition, the owner of a network will agree over financing ways of investment in the transmission network, established by the transmission and system operator and approved by NRAE, and will offer guarantees for prospective extensions of the network, with the exception of investment for which they agreed to be financed by any interested party, including the transmission and system operator.

142. Moreover, according to the Civil Code, they will bear responsibility with regard to the assets of the transmission network, with the exception of the events caused by the transmission and system operator. The Competition Council, in collaboration with NRAE, is responsible for monitoring whether the owner of the transmission network fulfils its obligations.[112]

§6. The Independence of the Owner of a Transmission Network

143. In case that the owner of the transmission network is part of a vertically integrated undertaking, it must be independent at least in the legal status, organization and decision process for other activities, that are not related to the transmission of electricity. The purpose of this measure is the practical implementation of the Directive 2009/72/CE: to guarantee the full independence of the network exploitation activities from the interests related to production and supply, as well as to prevent the exchange of confidential information.

144. To ensure the independence of the owner of a transmission network, the law provides for a series of minimal criteria which must be met:

- the persons who hold executive positions with the owner of a transmission network cannot be part of the structures of the integrated undertaking that is directly or indirectly responsible of the daily management of electricity production, distribution and supply activities;
- the persons who hold executive positions with the owner of a transmission network act independently from any market interest in fulfilling their work duties;

111. *See* Art. 39 of Electricity Law.
112. *See* Art. 37 of Electricity Law.

- the owner of the transmission network establishes a conformity programme which contains the measures taken to guarantee the exclusion of discriminatory practices and establishes the specific obligations imposed to the employees to achieve the independence objective;
- the owner of a transmission network appoints a 'conformity agent' who properly monitors the conformity programme and submits to NRAE in December each year a report regarding the taken measures, which will be published on the site of the transmission and system operator.

Likewise, the owner of the transmission network submits for approval, to NRAE, all draft agreements that are to be concluded with the transmission and system operator, including those referring to existent assets or those performed as a consequence of investments in the transmission network.[113]

§7. Interdictions Regarding the Transmission Network

145. For the protection of the transmission networks, the law provides for a series of interdictions which target both natural and legal persons. Thus, it is forbidden to build any kind of constructions within the safety area of the installations, without the approval of location from the transmission and system operator. In addition, it is forbidden to make diggings or set up plantations, without the approval of the transmission and system operator, as well as to deposit materials on passage ways, to throw objects or to intervene on electrical transmission networks by unauthorized persons. Furthermore, natural or legal persons must refrain from deteriorating constructions, fences or warning labels on the transmission installations as well as from limiting or preventing access to installations, by building constructions around them or in any other way.

113. *See* Art. 38, paras 2–3 of Electricity Law.

Chapter 6. Electricity Distribution

146. According to the law, electricity distribution is carried out by the distribution operator, legal person, license holder. The services provided by the distribution operators are for all users of electrical distribution networks, in non-discriminatory conditions, providing access to any applicant who meets the legal requirements. Similarly with the land on which transmission networks are placed, the lands on which there were distribution networks at the time when the Electricity Law came into force, were and remain in public property of the state, with the exception of the lands for which the distributor operators had already obtained ownership rights under the conditions of the law. Likewise, the distribution networks are separated from the production installations or from the electricity transmission networks and from the ones of final clients in the points of patrimonial delineation.[114]

147. The distribution networks are developed with the observance of economic efficiency, urban plans, property rights, environmental protection, persons' health and lives and energy saving, according to technical and safety norms included in the technical regulations in force.

148. The expenses for modifying the distribution installations as a result of the connection of new users or of the change in initial power characteristics of existing users, are to be paid entirely, according to NRAE regulations, Guidelines from 22 May 2013,[115] based on objective criteria.

§1. Distribution Operator

149. The distribution operator has, according to the law, the following main obligations: to exploit, to upgrade, to rehabilitate and to develop electrical distribution networks; to ensure the transit of the electricity through electrical distribution networks, upon the request and by informing the transmission and system operator; for those areas of the country where there is not enough capacity through transmission networks, in order to evacuate power from power plants, or in order to connect to the power system of a neighbouring country, to perform works for the development of electrical distribution networks through optimum development programmes; to provide the operative management according to the distribution license; to provide in non-discriminatory conditions information regarding their own activities, necessary for the network's users (without disclosing confidential commercial information obtained during its activity); to submit for approval to the transmission and system operator the programme of repairs and planned maintenance in installations with the nominal line of 110 kV and to monitor the safety in

114. *See* Art. 44, paras 1–5 of Electricity Law.

115. Guidelines concerning the regulation of conditions for taking over energy distribution facilities of electricity of 22 May 2013, were published in the Official Gazette of Romania, Part I, No. 308/ 29.05.2013.

operating the electrical distribution networks, as well as the performance indicators.[116] The distribution operator must also execute works of implementation/upgrading/rehabilitation/modernization of electrical networks in order to increase the quality level of the electricity distribution service.

150. With respect to electricity produced from renewable energy sources, the law states that the distribution operators are obliged to ensure the distribution of electricity, making sure that the electricity network is safe and reliable.[117] In the cases where the distribution operator is taking significant measures to limit the use of renewable energy sources, for the purpose of ensuring the safety of the national electric system and energy supply, then the system operators must inform NRAE of these measures and of the solutions intended for preventing inadequate limitations.[118]

151. In order to carry out any of the above-mentioned duties, the distribution operator is legally allowed to interrupt the operation of installations for performing the planned repair and maintenance works. In addition, the distribution operator may also interrupt the electricity distribution in case of danger to people's life, health or integrity of physical assets, to prevent or limit damage to energetic equipment, or to execute operations or works which cannot be carried out without interruptions.[119] The interruption may be only temporary – for the time necessary for performing the works or repairs – and with the prior notice to their related electrical distribution network users.[120] However, the operator is responsible for any damage caused to users due to its fault.

152. The distribution operator can take part in the wholesale electricity market, but only for the transaction necessary for their own technological consumption, corresponding to the electrical distribution networks and for their own consumption sites. In addition, in order to carry out works for quality level increase, the distribution operator has the right to place technical equipment on public domain, within the build-up areas, underground and on surface, under the condition to not interfere with public circulation and by obeying the regulations in force regarding placement of urban networks.

§2. The Obligations of the Owner of a Distribution Network

153. With the entering in force of the Electricity Law in 2012, any person who owns in property or has in use their own electrical distribution network bears the obligation to provide free access to it, and thus, if they are incapable to provide such an universal service for the users connected to that network, the network may be

116. *See* Art. 46, para. 1 of Electricity Law.
117. Article 9, para. 1 of Law No. 220/2008.
118. Article 9, para. 5 of Law No. 220/2008.
119. *See* Art. 47, Para of Electricity Law.
120. *See* Art. 45, para. 2 and Art. 47, para. 2 of Electricity Law.

taken over, with NRAE's approval and with suitable compensation, by the distribution operator, holding a license, from the area where the network is placed.

154. Considering that the owner of the distribution network is and remains separated from the distribution operator, the law attributed the former the following obligations: to cooperate with the distribution operator in order to fulfil their duties; to cooperate with NRAE and the distribution operator in establishing the investment in the distribution network, the corresponding financing and the transfer of the distribution service to another distribution operator; to facilitate the financing of prospective network extension, with the exception of those for which they agreed to be financed by any interested party, including the distribution operator.

155. The network owner will, however, be held responsible with regard to the assets of the distribution network, according to the general rules established by the Civil Code, with the exception of the responsibility regarding the distribution operator's attributions.

§3. Separation Between Distribution and Supply Activities

156. According to the law, if the distribution operator is part of a vertically integrated undertaking, it must be independent in their legal status, organization and decision process, for the activities that are unrelated with the distribution. However, there is no obligation to separate ownership over the assets belonging to the distribution operator from those belonging to the vertically integrated undertaking.

157. In order to evaluate and ensure the independence of the organization and of the decision process of the distribution operator, the law provides for a series of minimal criteria which must be met, somehow similar with those provided for the separation of the owner of a transmission network:[121] the persons responsible for the distribution operator's management cannot take part in the structures of the vertically integrated undertaking responsible, directly or indirectly, with the daily operation of the production, transmission and supply activities; the necessary measures must be taken so that the persons responsible with the distribution operator's management have independence in action. The distribution operator has the right to

121. Vertically integrated undertakings who supply services for less than 100,000 connected customers or who serve smaller systems, isolated, have no obligation to implement measures to ensure the independence of the distribution operator. The exception is explicitly stated in Art. 48, para. 3 of Electricity Law.

take decisions effectively, independent of the vertically integrated undertaking[122] with regard to the necessary assets for the operation, preservation or development of the network.[123]

158. At the same time the distribution operator has the obligation to exclude discriminatory practices, to establish specific obligations imposed on the employees in order to achieve the independence objective through a conformity programme and to designate a person or a body – a conformity agent – to monitor the programme and to submit to NRAE, annually, a report containing the measures that have been taken. In its turn, the conformity agent is completely independent and has access to all the information of the distribution operator and their affiliated undertakings, necessary in order to fulfil their obligations.[124]

159. In close connection with the separation of the distribution operator from the activities of the vertically integrated undertaking whose part they are, the distribution operator has the obligation: to keep the confidentiality of the commercial information obtained while performing its activity; to prevent discriminatory disclosure of information regarding its own activities, which may be commercially beneficial; and to not create confusion regarding their separate identity from the undertaking who provides the supply service of electricity within the vertically integrated undertaking. NRAE is responsible for monitoring the activity of the distribution operator who is part of a vertically integrated undertaking and for taking measures to prevent or eliminate potential anti-competition offences.[125]

§4. Measures for Protecting the Distribution Networks

160. In order to protect the distribution networks, the law provides for a series of interdictions for both natural and legal persons who must refrain from building any constructions, making diggings of any kind or setting up plantations within the safety areas of electrical distribution networks without the approval of the distributor operator. In addition, it is forbidden to deposit materials on passage ways and in the installations' protection and safety areas, without the approval of the distribution operator, to throw objects of any kind on electrical distribution networks or to intervene on them in any other way, to deteriorate constructions, fences or identification/warning labels related to the networks, or to limit or prevent in any way the access of the distribution operator to the installations.

---

122. It is explicitly forbidden by law for the parent company to give any kind of instructions regarding the distribution activity or to take individual decisions related to the construction or the upgrade of power distribution facilities which do not exceed the terms of the approved financial situations or any equivalent instrument. *See* Art. 48, para. 3, letter c) of Electricity Law.
123. In order to fulfil this obligation, the parent company must put at the distribution operator's disposal the necessary resources, including human, technical and financial resources. *See* Art. 48, para. 6 of Electricity Law.
124. *See* Art. 48, paras 1–2 of Electricity Law.
125. *See* Art. 48, paras 4, 5, 7 and 8 of Electricity Law.

§5. Isolated Distribution System

161. By definition, the isolated distribution system is the system through which electricity is distributed in an industrial, commercial or common services area, geographically limited and by which electricity is not supplied to household customers, if for technical or specific security reasons, the activities or the production process carried out by the users of that system are integrated or that respective system distributes electricity, mainly to the owner of the distribution system, its operator or their economically affiliated operators.

162. The isolated distribution system operator is under no obligation to purchase the electricity necessary for their own consumption based on transparent, non-discriminatory procedures, and is not subjected to the obligation that the prices for the connection to the network or the methodologies for its calculation be approved prior to coming into force. These exceptions also remain valid if the area served by an isolated distribution system includes household customers, under the condition that their annual consumption be lower than 5% from the annual consumption of the isolated power system and if these final clients have labour relations or are in any form of association with the owner of that respective distribution system. Nonetheless, the tariffs can be approved and revised by NRAE, at the request of an isolated distribution user.[126]

§6. Electrification of Localities and Expansion of Distribution Networks

163. The distribution operator is obligated, under the performance of the concession agreement and at the local or central public administration's request, to ensure the development and the financing of the distribution network for the electrification of the localities or for the expansion of the distribution network in the area covered by the concession agreement, respectively by the license, which they own. If the requested investment is not economically valid for the distribution operator, they have the right to propose co-financing to the requester, with prior information of NRAE. The evaluation of the financing conditions of such evaluations is to be done according to NRAE Guidelines of 23 October 2013.[127]

126. *See* Art. 50, paras 2–4 of Electricity Law.

127. Guidelines for evaluating the financing terms of investments regarding the electrification of localities or for extending the electricity distribution networks of 23 Oct. 2013 was published in the Official Gazette of Romania, Part I, No. 666/30.10.2013.

Chapter 7. The Electricity Market

164. The electricity market consists of the regulated market and the competitive market. The increase in percentage of the latter is to be done gradually, by ensuring the access to the market to as many participants, producers, suppliers or final clients as possible. With regard to the energy transactions, they can be wholesale or retail.

165. For solving the disputes between the participants in the wholesale and retail market, the Electricity Law established the Dispute Resolution Commission, which is formed by five members, appointed by NRAE's president's decision, for a three-year term. The members of the commission are NRAE specialists with a seniority of minimum eight years in the electricity field and perform their activity under the organizational and operational regulations approved by NRAE's president's decision.[128]

§1. The Participants in the Electricity Market: General Obligations

166. The participants in the electricity market and the associated operational structures are: the producer, the transmission and system operator, the electricity market operator, the distribution operator, the supplier and the customer.

167. The participants in the electricity market must *comply with its operation regulations* and must *make the payments* for the electricity and the services provided to them in due time, according to the agreements concluded between the parties. The rules for the operation of the energy market are established by the orders, decisions and approvals issued by the competent authority.[129]

168. The participants in the market have, as well, the obligation *to support the payment of the imbalance they generate* on the electricity market, with the exception of the imbalance for which redistribution rules are established by the competent authority or are explicitly provided by the law in force. At the same time, they *must strictly comply with the provisions of the licenses and regulations issued by NRAE*. With regard to the payment obligations related to the imbalances, they are to be paid within one month since the conclusion of the transaction, based on the measured quantity of electricity that represents the object of the transaction.[130]

169. On their side, the customers *must pay in due time the value of the invoices* that correspond to the electricity and services provided to them, as per the provisions of the agreements they concluded. At the same time, the customers can be obligated *to post a financial guarantee* in order to prevent the risk of failure to pay the electricity on the market.[131]

128. *See* Art. 20 of Electricity Law.
129. *See* Art. 21, para. 1 of Electricity Law.
130. *See* Art. 21, para. 3 of Electricity Law.
131. *See* Art. 21, para. 4 of Electricity Law.

170. The participants on the electricity market have a series of *obligations to inform* as well. Thus, they have to notify the transmission and system operator related to: imports, exports and transits on the transaction intervals with external partners, for each border, with regard to all the mutual contracted net quantities. In addition, they have the obligation to notify the electricity and balancing market operator with regard to the changes of the electricity quantities traded, corresponding to the electricity sale-purchase agreements concluded.[132]

§2. Electricity Market Operator

171. The electricity market operator is a legal person, license holder, which organizes and manages the electricity centralized market, with the exception of the balancing one, in order to carry out short, medium and long-term transactions of electricity, according to NRAE regulations. Like the transmission and system operator, the electricity market operator is forbidden to disclose information related to electricity transitions which they own and which they achieved during their activity, except as in the conditions of the law.

§3. The Operation of the Electricity Regulated Market

172. Initially, the electricity regulated market was created in order to ensure the electricity supply to the final customers who, as of the date when the Electricity Law entered into force, did not make use of their eligibility right, and to the household and non-household customers with a medium number of employees lower than fifty and a turnover or with the total book value of the assets below EUR 10 million.[133] However, currently, the regulated market aims only at the final customers who did not make use of their eligibility right and at the residential customers.[134] The supply of electricity towards these two categories is made based on regulated agreements, under the provisions of the Guidelines from 31 August 2016.[135]

173. The regulated prices are gradually eliminated, based on a calendar which is differentiated according to the affected customers and issued so that the possible negative consequences triggered by the elimination of the regulated prices shall affect as little as possible the protected categories.[136]

---

132. *See* Art. 21, paras 5–7 of Electricity Law.
133. *See* the corroborated provisions of Art. 22, para. 1 and Art. 55, para. 1 of Electricity Law.
134. According to Art. 21, para. 1, second thesis, of the Electricity Law the regulated market ensures the electricity supply to the household customers until 31 Dec. 2017.
135. The new Guidelines for monitoring the regulated electricity market were approved by NRAE's Order 42/24.08.2016, and were published in the Official Gazette of Romania No 673/31,08.2016, thus amending and replacing the NRAE's Guidelines of 9 Nov. 2013 on monitoring the regulated electricity market were published in the Official Gazette of Romania, Part I, No. 586/16.09.2013.
136. *See* Art. 22, para. 8 of Electricity Law.

§4. The Role of NRAE in the Operation of the Regulated Electricity Market

174. In order to ensure the operation of the regulated electricity market, the competent authority has the right to impose public service obligations to the producers, the transmission and system operator, distribution operators and last instance suppliers. According to Electricity Law,[137] the obligations of public service can target the safety, the quality and the price of the supplied electricity, the continuity in supply and energy efficiency. NRAE can impose to the last instance suppliers transparent procedures for the purchase of the electricity from the competitive market, if this is intended for household customers[138] and is to approve the guidelines for establishing costs or regulated income[139] for the purchase or sale of the electricity intended for customers on the regulated market.[140] In addition, they are able to approve and publish the prices and quantities from the agreements concluded between the producers and the suppliers of the household customers, as well as the prices used by the last instance suppliers. NRAE will notify and publish the prices proposed by the last instance suppliers for the electricity supplied to the residential customers, after the elimination of the regulated prices. In order to ensure the NES safety, the competent authority is able to approve the prices and quantities from the purchase and supply agreements for the technological system services.

175. Within the monitoring activity, the competent authority will follow the effect of the regulated market over the competitive market and will take the necessary measures to avoid possible distortions, as well as to increase the transparency degree of the commercial transactions. In addition, it will annually organize a detailed evaluation process of the operation of the energy market, under the conditions of gradually eliminating the application of the regulated prices for the non-household customers that are still the beneficiaries of the regulated electricity market, on the basis of certain general minimum indicators established by the law, and will make a biannual report with regard to the prices regulated as public service obligations and to their impact on the electricity market.

§5. The Operation of the Competitive Market

176. The electricity transactions performed on the competitive market are done with transparency, publicly, centralized and discrimination free, either wholesale or

137. *See* Art. 16 para. 1 of Electricity Law. According to para. 2 of the same article, the public service obligations may be established by Government Decision or NRAE's regulations (in which case NRAE will notify the European Commission), by licenses and establishment authorizations.

138. According to Art. 22, para. 2, letter e) of Electricity Law NRAE approves the monitoring guidelines for the expenses generated by electricity purchase on the competitive market for residential customers.

139. These costs may be increased only with the approval of the competent authority, solely in exceptional situations, of force majeure or triggered by law changes. *See* related provisions of Art. 78, para. 2 of Electricity Law.

140. *See* Art. 78, para. 1 of Electricity Law.

retail, and the prices are formed based on supply and demand. With regard to the wholesale market, the suppliers sell electricity to the final customers through bilateral agreements, either with negotiated prices, or with standard offers.[141]

§6. The Role of NRAE in the Operation of the Competitive Market

177. NRAE has the right to suspend the operation of the competitive market in case of an unexpected crisis on the energy market and in case of threat to the safety or security of persons, machines or installations or of the system's integrity. The suspension takes place at the proposal of the transmission and system operators, addressed to NRAE as well as to the relevant ministry.[142]

178. Safety measures are applied by Government Decision, at the initiative of the relevant ministry, and must have minimum influence to the best operation of the internal European market and to be limited to the remedy of the crisis situation which generated them. In addition, they must be immediately communicated to the other Member States and to the European Commission. The obligation to inform lies within the duties of the relevant ministry.

§7. Limitation of Prices and Tariffs

179. In cases of major unbalance between supply and demand and obvious disruptions of the electricity market, at NRAE's proposal and with the approval of the Competition Council, it is possible to limit the excessive price or tariffs increase or to freeze them for a determined period of maximum six months, which can be extended successively for periods of maximum three months, for as long as the causes that determined this action last, either by establishing an upper limit of the price on a centralized market, or by setting a limit to the income from the regulated activity. In the latter case, the acknowledged and postponed prices will be entirely recovered according to a procedure issued by NRAE.

---

141. *See* Art. 23, paras 1–4 of Electricity Law.
142. *See* the corroborated provisions of Art. 23, para. 8 with those of Art. 24 of Electricity Law. The execution method of the right to suspend the competitive market operation is provided for in NRAE's Regulation concerning the organized framework for trading on the centralized market through double continuous negotiation of electricity bilateral agreements of 12 Jul. 2013, published in the Official Gazette of Romania, Part I, No. 467/29.07.2013.

Chapter 8. The Green Certificates Market

180. The Green Certificates Market is a competitive market, separated from the electricity market, where Green Certificates are traded.[143] It has two components: a centralized market and a bilateral contracts market.[144] Both producers of electricity from renewable energy sources and suppliers are able to trade Green Certificates on each of the two markets. However, although the two markets are separated and function independently from each other, they both belong to the support scheme implemented by Law No. 220/2008. The trading framework is ensured by OPCOM, the electricity market operator, according to the NRAE regulations.[145]

Starting with 2015, electricity suppliers and producers are obliged to purchases every trimester a number of Green Certificates equal with the result of multiplying the value of the annual mandatory quota of purchase of Green Certificates for that year, and the quantity of electricity purchased and used for the final consumption of the suppliers', respectively, producers' activity (as well as the one invoiced to final consumers)[146], or the electricity used by a producer for supplying electricity to the consumers connected by direct lines to the power plant.[147] Suppliers of electricity and producers who are obliged to purchase Green Certificates and who achieve the mandatory annual quota for the acquisition of Green Certificates for the previous quarter in a proportion lower than 90% of the mandatory annual quota, are sanctioned by a fine equal with the product of the maximum value of the Green Certificates established annually by NRAE, according to the legal provisions and the number of Green certificates not purchased, corresponding to the difference between 90% of the mandatory annual quota and the share achieved at the level of each trimester.[148]

§1. Role of the Market Operator (OPCOM) in Administering the Green Certificates Market

181. OPCOM has the following competences: to draw up and update the Register of Green Certificates and the Register of Participants to the Green Certificates Market; to record in the Register of Participants the participants to the Green Certificates market on the basis of convention for participation; to record in the accounts of renewable energy producers the numeric codes of Green Certificates

143. Article 5, para. 1 of NRAE's Regulation from 1 Apr. 2015 concerning the organization and functioning of the Green Certificates market (hereinafter Regulation of Green Certificates Market) published in the Official Gazette of Romania, Part I, No. 287/28.04.2015.
144. Article 10, para. 1 of Law No. 220/2008 corroborated with Art. 5, para. 2 of Regulation of Green Certificates Market.
145. Article 10, para. 2 of Law No. 220/2008.
146. Excluding their own technological consumption.
147. Article 8, paras 1, 2 and 2^1 of Law No. 220/2008, as amended by Law No. 122/2015.
148. Article 12, para 2^3 of Law No. 220/2008, as amended by Law No. 122/2015.

issued by the Transmission and System Operator; to administer transparently, objectively and non-discriminatorily the transactions which take place on the two markets for Green Certificates; to ensure the takeover, validation and processing of sale-purchase offers for Green Certificates; to set up and upload on its own website the number of traded Green Certificates; to establish the rights and obligations of payment of the participants to the market; to ensure publication of announcements concerning the organization of public tenders and to organize public tenders on the centralized market for Green Certificates; to publish the results of the public tender, by mentioning the contracting parties and the sale price; and to transfer the traded Green Certificates from the seller's account to the buyer's account for the transactions concluded on the Green Certificates Market.[149]

182. The solution adopted by the Romanian legislator was that until reaching the national targets concerning the share of electricity produced from renewable energy sources in the gross final consumption of electricity, the trading of this type of electricity takes place solely for covering the final gross consumption of Romania. The said electricity is to be accompanied by certificates of origin, as provided by the regulations into force.[150]

§2. The Value of Green Certificates

183. The value of Green Certificates on the centralized market is the closing price of the market, set at the intersection between certified supply and demand curves. However, the closing price must lie somewhere between the minimum and maximum legal limits.[151] The price and the participants to the trade are established by OPCOM, but the financial transactions take place directly between the participants (producers and suppliers) without any financial transactions through the same operator. The trading period starts on first of March and ends at the end of February, the following year. Starting with 2011, the value of Green certificates is updated yearly by NRAE with the value of the average annual inflation rate, for the euro zone of the EU, as it is officially communicated by EUROSTAT.[152]

184. Thus, the Romanian Green Certificates system provided for a centralized market with prices set through competitive mechanisms, however set between limited minimum and maximum trading values per certificate. The same limits were imposed for the bilateral contracts market. The reasons were twofold: on the one hand, the minimum price was ensuring the bankability of the project concerning energy from renewable sources, while, on the other hand, since the costs are supported by final consumers, a maximum price was ensuring that the population and industrial customers will not be overburdened by the costs. The first measure proved

149. For the complete list of obligations and competences *see*: Art. 25 of Regulation on Green Certificates Market.
150. Article 10, para. 3 of Law No. 220/2008.
151. The prices for 2008–2025 are set in Art. 11 of Law No. 220/2008.
152. Article 12, para. 3 of Law No. 220/2008.

to be successful and large investments in renewable energy sources were undertaken in Romania. However, reaching the national target and the low volatility of Green Certificates prices (which settled towards the maximum limit) cause economic hardship on large energy consumers and caused, as already stated, the intervention of the Government and the amendment of the law.

185. Thus, for the period between 1 July 2013 and 31 March 2017, the trading of a number of Green certificates was postponed for each MWh produced and delivered by producers.[153] In other words, the Romanian legislator did not amend the price of the certificates but the mandatory number of certificates that was to be recovered.[154] The recovery is to take place in tranches between 2017 and 2020[155] but the high degree of uncertainty concerning the actual recovery of these certificates has caused many investors to abandon their ongoing or planned projects in Romania.

§3. The Bilateral Contracts Market of Green Certificates

186. On the bilateral contract market of Green Certificates, the participants are free to engage in bilateral transactions with Green Certificates. However, now the market consists of a centralized market for bilateral contracts concluded upon a public tender and a market for contracts concluded through direct negotiation between producers of electricity from renewable sources which hold groups or power plants which benefit from the support scheme, with a power of 1 MW per producer at the most and 2 MW for cogeneration of high efficiency based on biomass, and the suppliers of electricity for final customers. Participation in the bilateral contracts market is voluntary.[156]

I. The Functioning of the Centralized Market for Bilateral Contracts Concluded as a Result of a Public Tender

187. The scope of centralized market of bilateral contracts concluded as a result of a public tender is to ensure a transparent and non-discriminatory environment for trading Green Certificates among the participants, by setting public tenders[157] for purchasing Green Certificates for a determined period of time, at a transparent price, resulting from the balance between supply and demand.[158] Any transactions concluded on the centralized market create an obligation for winning participants of

153. Article 6, para. 2^1 of Law No. 220/2008.
154. The mandatory quota was set in Art. 6, para. 2 of Law No. 220/2008, while the new quantities are set in Art. 2^1 of Law No. 220/2008.
155. Article 6, para. 2^2 of Law No. 220/2008.
156. Articles 26 and 27 of Regulation of Green Certificates Market.
157. The provisions governing the organization of public tenders are contained in Arts 39–43 of Regulation on Green Certificates Market.
158. Article 29, para. 1 of Regulation on Green Certificates Market.

concluding the proposed and published contract.[159] The prices must remain within the maximum and minimum limits set by the law.[160]

II. Rights and Obligations of the Seller

188. The seller has the following rights: to send to OPCOM solicitations for organizing public tenders; to submit to OPCOM reply offers to published initiating offers[161] and participate in public tenders; to receive confirmation of transaction from OPCOM; to check, whenever he deems necessary the information from the Register concerning the situation of the Green Certificates held over transactions performed within a certain period of time; to solicit the correction of any inaccuracies identified in his account from the Register.[162]

189. The seller must also comply with a set of obligations, such as: to register himself as required by the law;[163] to conclude the contract in the published form and within the deadline established by the Regulation; to communicate to OPCOM the information concerning the transactions performed through the bilateral contracts concluded upon a public tender (the number of Green Certificates and their numeric codes) including the confirmations related to the performance of payment between the parties, within five calendar days; to send to the market operator copies of the bilateral contracts within three days from the transmission by the market operator of the transaction confirmation; and to invoice the value of the Green Certificates sold through the bilateral market, while complying with the terms and conditions set in the contract; and to pay to the market operator the related tariffs perceived.[164]

III. Rights and Obligations of the Buyer

190. The buyer enjoys the following rights: to submit to OPCOM solicitations to organize a public tender and offers to be published for trading; to submit to the market operator reply-offers to initiating offers and to participate in public tenders; to receive confirmations of its own transactions; to sell to other buyers the Green Certificates purchased over its annual mandatory quota, within the price limits established by the law; to use the Green Certificates purchased in a supplementary year over the mandatory quota for fulfilling the quota set for the next year, if these certificates maintain their validity; and to consult, anytime he deems necessary, the

159. Article 31 of Regulation on Green Certificates Market.
160. Article 32 of Regulation on Green Certificates Market.
161. The content of initiating offers is set in Art. 38 of Regulation on Green Certificates Market.
162. Article 33, para. 1 of Regulation on the Green Certificates Market.
163. This obligation derives from Art. 13 of the Regulation on the Green Certificates Market which states that all undertakings must register as participants if they wish to trade on the Green Certificates Market.
164. Article 33, para. 2 of Regulation on the Green Certificates Market.

information in the Register concerning the situation of the Green Certificates held over the transactions performed within a certain period of time.[165]

191. With respect to the obligations that the buyer must comply with, the law mentions: registering with the Green Certificates market,[166] as provided by the law; to conclude the bilateral contract in the published form and at the deadline established by the Regulation; and to pay the price of the purchased Green Certificates, in full compliance with the rights and obligations set out in the contract and submit the evidence of payment to the market operator; and to pay to the market operator the related tariffs.[167]

IV. The Functioning of the Bilateral Contracts Market Concluded Through Direct Negotiation

192. On this market, participants are free to engage in bilateral transactions of Green Certificates, at negotiated prices, for determined periods of time. Prices cannot be lower or higher than those established by the law.[168]

193. Among the rights and obligations mentioned by the Regulation for both sellers and buyers, one can find the right to conclude one or more bilateral contracts. The rest of the rights and obligations are similar to those of sellers and buyers involved in the bilateral contracts concluded upon a public tender, less the rights and obligations related to the initiating offers and public tenders.

§4. The Functioning of the Centralized Green Certificates Market

194. The centralized market's purpose is to make available for participants to the Green Certificates market a competitive framework for trading green certificates transparently, non-discriminatory, with increased liquidity and to offer them a reference price for concluding bilateral contracts according to predefined rules. Participation in the centralized market is voluntary and available for all participants in the Green Certificates market. The centralized market is supervised and administered by OPCOM, like all the other markets concerning Green Certificates.[169]

195. The transactions concluded on the centralized market are concluded at least once per month, on the basis of supply and demand requests of the participants to this market. The transactions concluded here create an obligation for the participant involved to sell, respectively, purchase the Green certificate.[170]

165. Article 34, para. 1 of Regulation on Green Certificates Market.
166. Article 13 of Regulation on Green Certificates Market.
167. Article 34, para. 2 of Regulation on Green Certificates Market.
168. Article 44, paras 1 ad 3 of Regulation on Green Certificates Market.
169. Articles 47–49 of Regulation on Green Certificates Market.
170. Articles 52 and 51 of Regulation on Green Certificates Market.

196. Rights and obligations of participants correspond with those described above for participants in other Green Certificates markets, however, in this case they will be entitled to submit for each trading session offers to sell or purchase Green Certificates, which will have the value of a firm commitment. The submission, the format and the frame content of the offers for the centralized market are set by the market operator.[171]

197. By law, no transaction can take place on the centralized market if: the curve of the supply and the one of demand do not meet and the smallest price for the aggregated sale offer is higher than the highest aggregated price of the purchase offer; if the supply is zero, if the demand is zero and if all the offers for the trading session are annulled.[172]

§5. SANCTIONS

198. Trading Green Certificates at prices outside the minimum and maximum limits set by the law constitutes a misdemeanour and it is sanctioned with a fine.[173] The amount of the fine varies depending on whether the person engaged in such trading is a natural or a legal person, the fines for legal persons being much higher.[174] The finding and sanctioning lies within the competences of NRAE's representatives.[175]

199. Since the law established a mandatory yearly quota of Green Certificates to be purchased, suppliers must prove annually to NRAE that they have complied with their obligation. In addition, starting with 2017 suppliers must also prove that they have purchased each trimester at least 90% of the number of Green Certificates as per provisions of the law. Where suppliers are in breach, NRAE imposes a penalty for each Green Certificate which was not purchased, in full accordance with the provisions of Law No. 220/2008.[176] These penalties are collected by OPCOM and transferred to the environment fund.[177]

---

171. Articles 53 and 54 corroborated with Art. 56 of Regulation on Green Certificates Market.
172. Article 59 of Regulation on Green Certificates Market.
173. Article 30, para. 1, letter c) of Law No. 220/2008.
174. Article 30, para. 2 of Law No. 220/2008.
175. Article 30, para. 4 of Law No. 220/2008.
176. Article 8, para. 1 corroborated with Art. 30, para. 1, letter b), letter b)[1] and para. 3 of Law No. 220/2008. Currently, this misdemeanour is sanctioned with a fine of RON 1.000 to RON 10.000, in the case of a natural person and RON 10.000 to RON 100.000 in the case of legal person, as per the amendments brought by GEO No 24/2017 concerning the amendment and completion of Law No. 220/2008 for the establishment of a support scheme for production of electricity from renewable source and for the amendment of certain normatives acts, published in the Official Gazette of Romania, Par I No 224/31.03.2017 (hereinafter GEO No 24/2017).
177. These penalties are revenues for the environment fund and are used for supporting the production of energy from renewable sources by natural persons who invest in producing electricity through installations with a capacity up until 100 kW.

§6. Resolution of Disputes Arising on the Green Certificates Market

200. Any dispute arising may be amiably settled among the involved parties, within fifteen days from its occurrence.[178] Where an agreement cannot be reached within the legal term, then the parties may seek resolution in accordance with the legal norms into force. The previous procedure provided for International Commercial Arbitration with the Tribunal of Romanian Chamber of Commerce but the new one did not maintain this requirement. However, nothing precludes parties of bilateral contracts, to mutually agree for arbitration as resolution method of their disputes.

178. Article 73 of Regulation on Green Certificates Market.

Chapter 9. Electricity Supply

201. Electricity supply is defined as the activity whereby the legal person license holder trades electricity to its customers, based on an agreement concluded according to the regulations in force. The electricity shall be invoiced according to NRAE's regulations. The consumption and the supply of electricity in absence of an agreement is forbidden by law, but NRAE has the right to approve of exception situations, which are the object of different regulations.

§1. Direct Electricity Lines

202. The producers and the suppliers of electricity can supply to their own office, subsidiaries or eligible customers through direct lines. The implementation of direct electricity lines, the access and tariffs for access to them is regulated by the competent authority. Likewise, their implementation must be done with the obeying of the technical safety norms included in the technical regulations in force. The condition for the supply by direct lines is the absence of a technically and economically reasonable offer for the access to the electrical network of public interest and must ensure the conditions with regard to consumers' protection are followed. The owner of a direct electricity line has the obligation to ensure the access to the direct line according to NRAE's regulations.[179]

§2. Connection Agreement

203. The connection agreement is an agreement under regulation by NRAE. The transmission and system operator or the distribution operator has the obligation, at the written request of a network user, to communicate in writing, within thirty days, the technical and economical connection conditions and to cooperate with the applicant in view of choosing the most convenient connection solution. However, the transmission and system operator as well as the distribution operator have the right to deny the renewal of the connection approval to an applicant and/or the conclusion of transmission or distribution agreement for a pre-existent consumption site, if for that consumption point there is already an agreement for electricity supply ongoing with another final client.

204. With regard to delineating the installations' area, the network operator and its user will consider: the installation's top, the property's limit over the land (so that the electricity networks placed on public land is owned by the network operator and to avoid as much as possible to place the installations of the network operator on the user's land) and the prospective connection of new users to the network. Needless to say, if the installations of the network operator were on the property of a user, connecting new customers could be made difficult by the latter's refusal.

179. *See* Art. 25, paras 4–10 of Electricity Law.

§3. Last Instance Suppliers: Rights and Obligations

205. Last instance suppliers are appointed by NRAE out of the already existing suppliers on the power market, through competition mechanisms implemented by Order 119/2017,[180] which states the selection modalities and criteria, for each category of final clients to whom they are addressed. The last instance suppliers' tariffs include the entire electricity purchase costs, the associated costs, as well as a reasonable profit share.[181]

206. The main obligation of last instance suppliers is to provide an universal[182] service to household and non-household customers with an average registered employee number lower than fifty and an annual turnover with a total value of assets on the balance sheet under EUR 10 million. In addition, the last instance suppliers have the obligation to supply electricity to final clients whose supplier is about to have the operation license withdrawn during the operation or in any other situation identified by NRAE as affecting the electricity supply to final clients.[183] The transfer of a client found in such situation is done without concluding a supply agreement, at prices and tariffs decided or approved under NRAE's regulations.[184] The sole exception from this latter obligation targets the final clients who have been disconnected for stealing or failure to pay for electricity.

207. Last instance suppliers have the obligation to supply power in quality conditions and at reasonable, transparent comparable and non-discriminatory prices, according to NRAE's regulations and by complying with the legal provisions regarding the regulated market, not only to the household and non-housed customers mentioned above, but also to the final clients who, at the date when the Electricity Law came into force, did not exercise their eligibility right.

208. The supplier must provide the household customers with unique contact points for their information regarding their rights, the legislation in force and the ways for resolving possible disputes or complaints. Moreover, the household customers must have access to several payment methods for the supplied power and the possibility to choose any of them. In addition, the supplier must label the electricity according to NRAE's regulations and inform periodically the final clients regarding the structure, the origin, and the impact on the environment of the supplied power.

180. Order No. 119/2017, concerning the designation of last instance suppliers, was published in the Official Gazette of Romania No. 1004/18.12.2017. It abrogated and replaced Order No. 35/2014 concerning the designation of last instance suppliers was published in the Official Gazette of Romania, Part I, No. 442/17.06.2014.
181. *See* Art. 79, para. 6 of Electricity Law.
182. Violation of this obligation constitutes misdemeanour according to Art. 93, para. 1, point 17 and is sanctioned with a fine according to Art. 93, para. 2, point 2, letter a) of Electricity Law.
183. *See* Art. 53, paras 1–2 of Electricity Law.
184. *See* Art. 56, paras 1 and 3 of Electricity Law.

§4. Contracting Obligation: Termination of the Supply Agreement

209. According to Article 52 paragraph 1 of Electricity Law, the electricity supply is performed only based on an agreement. Upon the written request of a final customer, the supplier has the obligation to communicate in writing, within no more than fifteen working days, an offer regarding the electricity supply conditions, which will include the supply prices, payment terms and the deadline for concluding the supply agreement. The agreement will be valid for an undetermined period, if the parties do not agree on another term.

210. The final customer has the right to terminate the concluded agreement at any time, with a prior notification of minim twenty-one days, provided he complies with all the other conditions of the agreement. The supplier can terminate the agreement only in the cases explicitly provided for by the law: if there is a final court order indicating the final customer's fraudulent consumption, if the invoices are not paid, and in any other cases provided for by the law in force.

211. In order to eliminate or reduce the risk of non-payment, the Romanian law gives the supplier the right to request its final customers a financial guarantee for a consumption period equivalent to maximum one year in case they identify certain actions meant to tamper with the measurement equipment or intent of fraudulent consumption by avoiding the measurement equipment. This measure is preventive and can be used only in the above-mentioned cases.[185] The method of implementation is provided for in the framework agreements of electricity supply and/or in regulations issued by NRAE. The refusal to constitute the requested guarantees is sanctioned with the right of the supplier to request the distribution operator to disconnect the user from the network.

§5. Liabilities of the Supplier

212. The supplier is liable for all damage caused to the consumer of their fault, based on the conditions established in the supply agreement. The supplier is entitled to recover the damages created to final customers or to themselves, from the distribution operator, the transmission and system operator or from the generator, provided that it is proven to have been created by the fault of any of them. The measure, aimed to protect the interests of the final customers, leads to the conclusion that the supplier will be liable for the damage caused by operators, and they will take action against them later on, which represents an exception from the general rules of civil liability. Nonetheless, in my opinion in case of litigation, the supplier will have the right to 'call as guarantor' the operator who caused the damage, according to the rules of the Code of Civil Procedure.

185. *See* the corroborated provisions of Art. 58, para. 5 and Art. 61, para. 7 of Energy Law.

§6. Liabilities of the Final Customers

213. The final customer is mainly liable to make the payment in due time for the electricity consumed at the prices and under the conditions decided by the supply agreement, lest they should be subjected to penalties, temporary interruption of electricity supply or even the termination of the supply agreement. In addition, the interruption of power supply can also be done if unauthorized persons have made interventions on the equipment, according to NRAE's regulations. Moreover, if the indications of the measurement equipment have been tampered with or the measurement equipment has been avoided for fraudulent consumption,[186] the final client may be obliged to offer a financial guarantee, and its refusal could lead to disconnection from the electricity network. The final customers with an approved power of a minimum of 1,000 kVA by the technical connection approval have the obligation to submit an hourly consumption schedule to the suppliers with whom they have contractual relations.[187]

§7. The Rights of the Final Customers

214. The final consumers of electricity are entitled to access to electrical networks of public interest and to consume the electricity; to request the supplier to modify and supplement the supply agreement and the annexes thereto or to initiate addendums to agreements when new elements appear or when they consider necessary to detail or supplement certain contractual clauses; to have access to measurement settlement sets; to request the supplier or the distribution operator to remedy the defects and malfunctions occurring in electrical networks; to request indemnification from the supplier, network operator or another customer, according to the contractual clauses, if damages occur from their fault or if power quality indicators are not met; and to associate as volunteer in order to represent the customers' interests, according to NRAE's regulations.

215. In addition, the final clients have the right to purchase electricity based on a bilateral agreement from a power supplier, irrespective of the EU state where they are registered, under the condition that they met the applicable commercial and balancing norms established for Romania; to switch the supplier, according to the procedure approved by NRAE;[188] to submit to NRAE for resolution the disputes arising from agreements concluded in the electricity field, to benefit from the services of a

186. The deterioration, modification with no right or the blockage of electricity measurement equipment, as well as the execution or use of clandestine installations, in order to directly connect to the network or to avoid measurement equipment represent infractions and are punished with imprisonment. For reference, *see* Art. 92 of Electricity Law.

187. *See* the corroborated provisions of Art. 59 and Art. 61 of Electricity Law.

188. According to Art. 62, para. 3 of Electricity Law, the procedure in discussion guarantees to the final customers that the requested change in supplier may be realized in term of three weeks since its initiation.

mediator in order to solve miscellaneous litigations occurred on the electricity market and to unilaterally terminate the agreement by following the procedure and the notice period.

216. According to the law, certain categories of final clients have the right to choose any of the regulated tariffs applicable to the categories to which they belong or to benefit of the universal service, under the conditions established by NRAE. This is the case of customers who still benefit from electricity insurance through the regulated power market.[189] The household customers have the right to choose any payment method which is provided by the supplier.

217. In addition to the right stated above, the big non-household customers, whose maximum connection power approved by the technical connection permit is established by NRAE's regulations, have the exclusive legal right to conclude agreements with more power suppliers at the same time, for one consumption site.

§8. Protection of Vulnerable Customers

218. According to the provisions of the law,[190] certain categories of customers are considered vulnerable. It refers to the household customers who, due to age, health or low income, are in risk of being socially marginalized and who benefit from social protection, including financially. In order to protect them, the law established certain facilities, including access to the electricity supply service and access to network. All types of facilities, with the exception of the financial ones, as well as the corresponding implementation method are to be regulated by NRAE. It is explicitly forbidden to disconnect from the electricity network vulnerable customers, even in power crisis situations.[191]

§9. Measurement of Electricity

219. The electricity supply to final clients on the electricity market is measured by measurement operators, by measurement sets, according to the Measurement Code issued by NRAE. The measurement operators may be, according to the law, the transmission and system operators, the producer, the distribution operator. The obligation to provide the operation service falls within the responsibilities of the transmission and system operator, respectively of the distribution operator for the users of that electrical network, but the service can be granted by concession to third parties under the condition of keeping the prices within the limits acknowledged by NRAE, and by following the applicable regulations they issue.

189. *See* Art. 63, para. 1 corroborated with the provisions of Art. 55 para. 1 and Art. 22, para. 1 of Electricity Law.
190. *See* Art. 3, point 16 of Electricity Law.
191. *See* Art. 64 of Electricity Law.

Part II. Regulatory Framework Concerning Gas

220. As emphasized in Romania's Energy Strategy for 2007–2020, almost 69% of Romania's national transport system of natural gas is old and exceeded its lifetime, this issue being one of the major deficiencies of the Romanian system together with the lack of financing sources compared to the actual investment needs of the infrastructure. Out of the entirety of measuring and regulating stations, about 27% have been in operation for over twenty-five years. The supply networks are characterized by an increased level of usage of both pipes and connectors, around 40% exceeding their lifetime.[192] However, the capacity for underground storage has been permanently developed, up to 4 billion cubic metre.

221. In Romania, 98% of the national production is realized by two companies, Romgaz and Petrom, while the remaining 2% is represented by other companies. Around 62.5% of the total national production is extracted in Mures County.[193]

222. The national market for national gas has been gradually opened starting with 2001, when the initial opening amounted to 10% of the total consumption for the year 2000 and in 2006 it reached an opening of 75%. The market liberalization continued and by 1 January 2007, the date of Romania's accession to the EU, it reached 100% for industrial consumers. At the present moment the market is liberalized for residential consumers as well, though not entirely.[194]

192. Romania's Energy Strategy for 2007–2020, pages 19 and 23, http://mmediu.ro/new/wp-content/uploads/2014/01/2011-11-07_evaluare_impact_planuri_strategiaenergeticaactualizata2011.pdf, last visited 23 Mar. 2018. Text available in Romanian only.

193. Romania's Energy Strategy for 2007–2020, page 21, http://mmediu.ro/new/wp-content/uploads/2014/01/2011-11-07_evaluare_impact_planuri_strategiaenergeticaactualizata2011.pdf, last visited 23 Mar. 2018. Text available in Romanian only.

194. Romania's Energy Strategy for 2007–2020, page 22, http://mmediu.ro/new/wp-content/uploads/2014/01/2011-11-07_evaluare_impact_planuri_strategiaenergeticaactualizata2011.pdf, last visited 23 Mar. 2018. Text available in Romanian only.

Chapter 1. General Introduction

223. The gas sector in Romania is mainly governed by the provisions of Law No. 123/2012 of concerning electricity and natural gas (hereinafter Natural Gas Law), with all further amendments. Title II of Law No. 123/2012: 'Natural Gas' transposes the provisions of Directive 2009/73/CE concerning the common rules for the internal market in natural gas[195] and repealed the Gas Law No. 351/2004.[196] It establishes the regulatory framework with respect to carrying out activities regarding production, transport, distribution, supply and storage of natural gas, the organization and functioning of the natural gas sector, access to the market as well as the criteria and procedures applicable for issuance of authorizations and/or licenses for the natural gas sector.[197]

224. According to Natural Gas Law, all activities within the gas sector are carried out for the following core objectives: ensuring durable development of national economy; ensuring continuity and safety in gas supply to customers; protecting the legitimate interests of final consumers of natural gas; promoting, stimulating and ensuring competition on natural gas market; harmonizing national legislation with EU legislation in the field; prices and regulated tariffs transparency within the gas sector; developing the natural gas sector while ensuring economic efficiency and environmental protection; promoting the production and use of new and renewable sources of gas, etc.[198]

§1. Energy Strategy and Policy in Natural Gas Sector

225. As already stated, the Romania's Energy Strategy is elaborated by the relevant ministry and approved by the government with prior consultation of NGOs, social partners and business sector's representatives.[199] Romania's Energy Strategy for 2016–2030 defines the medium and long-term objectives of the natural gas sector and the ways to achieve them, while ensuring a durable development of national economy. Given these objectives, the National Strategy emphasizes the importance of exploitation projects, both onshore and offshore, meant to reduce Romania's dependency on gas imports in the following decades;[200] the need to develop the infrastructure, mainly to link the Black Sea deposits with the National Transport

195. Last statement of Natural Gas Law.
196. Article 200, para. 2 of Natural Gas Law.
197. Article 98, para. 1 of Natural Gas Law. Paragraph 2 states that for the purposes of Natural Gas Law there is no distinction in regime between biogas, gas obtained from biomass or other types of gas, as long as it is technically possible to inject it or transport it through the transport/distribution system for natural gas and through the supply pipeline in complete safety.
198. For a full list of core objectives, *see* Art. 99 of Natural Gas Law.
199. Article 101, para. 1 of Natural Gas Law. Romania's Energy Strategy for 2016–2030, with a view towards the year 2050, is available in Romanian only at: http://www.mmediu.ro/app/webroot/uploads/files/2017-03-02_Strategia-Energetica-a-Romaniei-2016-2030.pdf, last accessed on 23 Mar. 2018.
200. Idem at 4.

System,[201] and to integrate the national market with the regional one, by finalizing the BRUA interconnector and the supply pipeline towards Republic of Moldova. The new strategy also recognizes the need to modernize and update the existing national infrastructure for transport, storage and distribution of national gas, to allow operations at high pressures, to avoid losses in the network and to increase operation flexibility. Last, but not least, another objective is to create a competitive market for natural gas, based on principles of transparency, liquidity, low level of concentration and competitive prices.[202]

226. The energy policy meant to implement the directives set in Romania's Energy Strategy for 2016–2030 is also elaborated by the relevant ministry, on the basis of the governing programme, for a medium term and taking into consideration the long-term probable evolutions, with prior consultation of NGOs, social partners and business sector's representatives, taking into account in particular: the establishment of an appropriate institutional framework, by setting up bodies and competent authorities for the implementation of this policy; ensuring the safety of natural gas supply; the forecasting of natural gas imports and exports; ensuring environmental protection; the increasing of economic and/or energetic efficiency of production, storage, transport, distribution and usage of natural gas.[203]

§2. The Relevant Ministry: Competences

227. The relevant ministry elaborates the policy in the natural gas sector and ensures that it is set into practice together with the Government and other special bodies of the public central administration. They also examine annually or whenever necessary the status of the policy's implementation.[204]

228. In fulfilling its role, the relevant ministry enjoys the following main competences:

- Implements the Government's energy policy.
- Elaborates programmes and plans for applying the Government's policy to the natural gas sector.
- Ensures the drafting of studies on the basis of which the investing priorities within the natural gas sector will be established.
- Draws up normative projects for the gas sector with consultation of all stakeholders.
- Exercises the role of concessionaire for the public service of natural gas distribution.

201. *Ibid.* at 5.
202. *Ibid.* at 5.
203. For a full list of factors to be considered by energy policy *see* Art. 101, para. 2 of Natural Gas Law.
204. Article 102 corroborated with Art. 101, para. 3 of Natural Gas Law.

– Supervises the enforcement and compliance of environmental protection measures.[205]

§3. NRAE: Competences Concerning the Natural Gas Sector

229. The main attribute of NRAE with respect to natural gas sector is to comply with and enforce all relevant and mandatory decisions of the Agency for the Cooperation of Energy Regulators (ACER) and the European Commission, of the Government, the resort ministry and all other specialized bodies of the central public administration, as the case may be. Additionally, NRAE cooperates and consults with all other regulatory bodies within the EU, by providing them and ACER with the necessary data for fulfilling their competences as of Directive 2009/73/CE. By law, NRAE is entitled to ask for ACER's approval concerning the compliance of any decision made by a regulatory body with the guidelines of Directive 2009/73/CE. NRAE may inform the European Commission in case it considers that any relevant decision for the cross-border commerce, made by another regulatory body, is not in compliance with the guidelines of Directive 2009/73/CE, within sixty days from the issuance of the aforementioned decision.[206]

230. NRAE approves by its President's Order and publishes technical guidelines[207] which establish the criteria for technical safety and minimum projection, performance and exploitation technical requirements for connecting to LNG terminals, storage facilities, other transport and distribution systems as well as direct main pipelines. Last but not least, NRAE supervises, through its representatives, the compliance with specifications, safety features and technical criteria of facilities and equipment used in the transport and gas distribution network approved in compliance with European legislation.[208]

231. As of NRAE's obligations, the agency must comply, within sixty days, with any decision of the European Commission which imposes the withdrawal or revocation of any NRAE decision and inform the European Commission thereof.[209]

§4. Regulatory Body: Power to Conduct Investigations

232. NRAE conducts investigations following its president's decision, through its own authorized personnel, either ex officio or following the complaint of a third party, directly affected by an alleged infringement of the law.[210]

205. For a complete list of powers and competences Art. 102 of Natural Gas Law.
206. Article 102^1, paras 1–4 of Natural Gas Law.
207. Article 102^1, para. 6. These guidelines must ensure systems' interoperability, must be objective and non-discriminatory and shall be notified to the European Commission in accordance with the provisions of Natural Gas Law.
208. Article 102^1, para. 7 of Natural Gas Law.
209. Article 102^1, para. 5 of Natural Gas Law.
210. Article 183 of the Natural Gas Law.

233. In order to conduct the investigation, NRAE is entitled to solicit all needed documents and information from the undertakings,[211] mentioning the legal and the scope of the request, and may set deadlines until the solicited data must be supplied.[212]

234. NRAE's authorized personnel may perform inspections solely on the basis of the authorization issued by the president of the competent authority and the judicial authorization[213] issued by the President of the Bucharest Court of Appeals,[214] and solely with full observance of the Regulation concerning the organization and performance of investigations.[215]

235. Throughout the procedure, the authorized personnel of NRAE enjoy the following rights: to enter the premises, lands and vehicles legally owned by the undertakings; to examine any documents, registries, financial, accounting or commercial documents concerning the undertaking's activity, notwithstanding the place where they are stored; to demand explanations concerning the facts or documents under investigation from any legal representative or employee of the undertaking and record their answers; to seize or obtain copies or excerpts of any document, register or document related to the activity of the undertaking; to seal any location or any document related to the investigation for the entire period necessary. NRAE is also entitled to conduct unannounced inspections and may solicit, within a reasonable term, any information or explanation related to the fulfilment of its investigation powers, both on the site or by summoning to its headquarter.[216]

§5. General Framework of Concessions Concerning Natural Gas Sector

236. According to Romanian law *all public goods* related to natural gas transport and storage objectives or systems *as well as* to natural gas transport, storage and distribution *services* constitute the object of concession to a national or foreign legal person.[217]

211. By Art. 188 public bodies as well as any other central or local administration institutions are obliged to allow access to the competent authority to documents, data and information they hold. However, the competent authority must comply with the status of state secret or work secret legally attributed to the documents, data or information in question.
212. Article 184 of the Natural Gas Law.
213. The judicial procedure takes place *in camera* without the subpoena of the parties.
214. For details *see* Art. 187 of Natural Gas Law.
215. Article 189 of Natural Gas Law.
216. Article 186 of Natural Gas Law.
217. Article 103 of Natural Gas Law.

Chapter 2. Licensing and Authorization Regime

237. In case foreign or national legal persons wish to set up new upstream supply pipelines, related to natural gas production, as well as new production objectives for biogas, bio-methane, LNG and CNG, or new systems for transport, storage and distribution of natural gas, they first must obtain from NRAE an authorization[218] thereof. However, for performance of any of the activities related to gas supply, transport, distribution, operation of upstream supply pipelines or LNG terminals, as well as administration of centralized gas markets, a license from NRAE[219] is required.[220]

238. Carrying out any of the regulated activities or of any works without holding the license or authorization required by the law or failure to observe the licenses' or authorizations' terms constitute misdemeanour and are sanctioned with a monetary fine.[221] The legal regime applicable is the general legislation governing misdemeanours, namely Government Ordinance No. 2/2001.[222]

239. The procedure for granting, modifying, suspending or withdrawing authorizations or licenses, together with the granting's terms and conditions are established by NRAE regulations.[223] Currently the regime of authorizations and licenses is governed by NRAE's Regulation concerning the granting of authorizations and licenses, approved by Order No. 34/31st of May 2013 of NRAE's president.[224]

§1. Procedures, Terms and Conditions for Granting Authorizations and Licenses

240. The procedure starts with an official request, based on the form provided by NRAE and accompanied by all documents required, made by a foreign or national undertaking. However, foreign undertakings are allowed to file such a request solely if they have already established a secondary headquarter in Romania. Operators under insolvency proceedings and those who had a license or authorization withdrawn in the past five years are precluded from filing.[225]

---

218. Article 6 of Regulation concerning granting of authorizations and licenses, approved by Order 34/2013.
219. Article 7 of Regulation concerning granting of authorizations and licenses, approved by Order 34/2013.
220. Article 118, paras 1 and 2 of Natural Gas Law.
221. *See* the combined provisions of Arts 194 and 195 of Natural Gas Law.
222. Article 198 of Natural Gas Law.
223. Article 118, para. 3 of Natural Gas Law.
224. Order 34/2013 was published in the Official Gazette of Romania, Part I, No. 427/15.07.2013 and was amended by Order 103/2014 of NRAE's President, published in the Official Gazette of Romania, Part I, No. 774/24.10.2014.
225. Article 8 of Regulation concerning granting of authorizations and licenses, approved by Order 34/2013.

241. After the analysis of submitted documents, NRAE may either notify the applicant seeking additional information[226] or issue a report recommending the approval or denial of the request[227] and submit it to NRAE's regulatory committee for approval.[228] In the case where additional information or clarifications were notified, the applicant must comply with NRAE's request within thirty days[229] or the request will be denied. In case the report makes a positive recommendation, it will also include the proposed validity term for the authorization or license.[230]

242. For both the analysis of the documentation and the grant/modification of the authorization or licenses, NRAE is entitled to perceive tariffs, approved by Order of its President. Hence, together with the required documents, applicants must also submit the proof of payment. Failure to pay the tariffs may result in NRAE's refusal to render a decision on the applicant's request.[231]

§2. Common and Specific Documents Required for All Types of Authorizations and Licenses

243. The request concerning the grant of an authorization or a license must be accompanied by a set of common documents. On the one hand, applicants must submit a certificate issued by the Trade Register not longer than thirty days before the date of submission to NRAE containing information regarding the identification of the applicant, its name, headquarter, status, the tax number, Trade Register number, social capital, legal representatives, registered activities and secondary offices in Romania. On the other hand, applicants must attach copies of IDs and CVs of their legal representatives and, in certain cases, of their manager responsible for technical issues. Foreign applicants are entitled to submit the requested documents issued by their competent authorities in certified translation. Where applicants request more than one authorization or license, a single copy of the aforementioned documents will suffice.[232]

244. Beside the common documents mentioned, applicant must submit a set of documents specific to the activity or service for which they require authorization or

226. Article 9, para. 1 of Regulation concerning granting of authorizations and licenses, approved by Order 34/2013.

227. Article 13, para. 1 of Regulation concerning granting of authorizations and licenses, approved by Order 34/2013.

228. Article 15, para. 1 of Regulation concerning granting of authorizations and licenses, approved by Order 34/2013.

229. Article 9, para. 4 of Regulation concerning granting of authorizations and licenses, approved by Order 34/2013.

230. Article 13, para. 2 of Regulation concerning granting of authorizations and licenses, approved by Order 34/2013. The proposed validity term is usually correlated with the contract on the basis of which the activity requiring authorization or license is performed.

231. Article 29 of Regulation concerning granting of authorizations and licenses, approved by Order 34/2013.

232. Article 10 of Regulation concerning granting of authorizations and licenses, approved by Order 34/2013.

license. The list of specific documents is detailed in Regulation concerning granting of authorizations and licenses, approved by Order 34/2013.[233]

§3. Approval or Refusal of Request for Granting Authorization or License

245. If the documentation is complete and NRAE has submitted a report recommending approval, the Regulating Committee of NRAE may decide, within thirty days from the submission, to approve the grant of the requested authorization or license. However, the grant is done through the Decision of the NRAE's President.[234]

246. The validity terms related to the granted authorizations and licenses contain, as the case may be, the following main provisions: the rights granted; the conditions, limitations and restrictions regarding the continuity and the functioning regime of the authorized objectives; the technical data concerning the authorized objective and the obligation of the grantees to supply NRAE with information concerning the activity performed in the natural gas sector as requested and established by NRAE.

247. Granted authorizations and licenses are registered in an electronic registry kept by NRAE. However, a copy is sent to the grantee and another is attached to NRAE's report which lies at the basis of the grant.[235]

248. The Decisions of the NRAE's president are published on the grantee's expense in the Official Gazette of Romania. They are also published on the webpage of the competent authority and may be subjected to judicial review in front of administrative courts, under the procedure established by Law No. 554/2004.[236]

249. In case of modification, denial, suspension or withdrawal of authorizations and/or licenses, the aforementioned provisions are also applicable with the sole difference that in the cases of denial, suspension or withdrawal, the Decisions of NRAE's president are published on NRAE's expense.[237]

233. Articles 11–12 of Regulation concerning granting of authorizations and licenses, approved by Order 34/2013.
234. Article 15 of Regulation concerning granting of authorizations and licenses, approved by Order 34/2013.
235. Articles 19 and 20 of Regulation concerning granting of authorizations and licenses, approved by Order 34/2013.
236. Article 21, paras 1, 3 and 4 of Regulation concerning granting of authorizations and licenses, approved by Order 34/2013.
237. Article 15, para. 2 and Art. 21, para. 2 of Regulation concerning granting of authorizations and licenses, approved by Order 34/2013.

250. According to the law, NRAE may decline to grant an authorization or a license where: the applicant does not fulfil the conditions set in the specific regulations; the applicant is under an insolvency proceeding or bankrupt; the applicant had an authorization or a license withdrawn by NRAE within the past five years for reasons attributed to it; an authorization or license has already been issued to another undertaking for building or exploiting distribution systems within a designated area. The reasons for denial must be objective and non-discriminatory.[238] The denying decision of the NRAE's president must be motivated and communicated to the applicant, who is entitled to contest it in administrative courts, and to the European Commission.[239]

§4. The Procedure for Modifying Authorizations and Licenses

251. Modification of authorizations and licenses may be decided either at the request of holders or by NRAE's own initiative, where the following situations occur: there is a need for updating the technical parameters mentioned; there is a change in the status of the holder;[240] there is a legislative change or change of the circumstances which were in place when the authorization or license was initially granted.[241]

252. The holder of an authorization may request modification either quarterly, where the objectives or systems are performed on the basis of technical documentations or projects which have not been included in the documentation on the basis of which the authorization was granted, or annually, in any other situations. However, the holders must submit explanations and justifications for the investments carried out outside the initial investment plan approved.[242] In such cases, beside the common documentation, the holder must submit also a specific set of documents.[243]

253. The holders of licenses may request modification either before commissioning the objective/systems, when they are not located in an area for which there was no specific license granted, or, annually, when newly commissioned objectives are located in areas for which a specific license has been granted.

238. Article 120, paras 1 and 2, first thesis of Natural Gas Law.

239. Article 120, para. 2, second thesis and para. 3 of Natural Gas Law.

240. In such cases, the holder is obliged to request modification within thirty days from the change in status. Where the modification is not possible, the holder may ask for the transfer of the authorization or the license to a different undertaking. However, the latter must submit a request and the necessary documentation to NRAE in order for the transfer to take place. Article 22, paras 3 and 4 of Regulation concerning granting of authorizations and licenses, approved by Order 34/2013.

241. Article 22, para. 1 of Regulation concerning granting of authorizations and licenses, approved by Order 34/2013.

242. Article 23 of Regulation concerning granting of authorizations and licenses, approved by Order 34/2013.

243. Article 25, para. 1 of Regulation concerning granting of authorizations and licenses, approved by Order 34/2013.

§5. The Procedure for Suspending or Withdrawing Authorizations or Licenses

254. In cases where NRAE has noticed on its own, by notification from third parties or from the holder, that the holder of the authorization or license does not meet the specific validity conditions for reasons which cannot be attributed to it, NRAE may:

- grant a term of maximum three months to ensure compliance, where the situation can be remedied;
- sanction the holder with a warning and temporary suspension of the authorization or license, for a maximum term of three months, if, at the expiration of the above-mentioned term the situation has not been remedied;
- withdraw the authorization or license, if, after the temporary suspension period, the situation has not been remedied.

In all the aforementioned cases, if the activity is performed on the basis of a concession agreement, NRAE will notify the concession grantor.[244]

255. Where the holder does not comply or improperly fulfils the validity conditions, NRAE may apply the same measures with the sole difference that in these cases, the warning sanction is replaced with a fine.[245] In both cases, the suspension or the withdrawal of the authorization or the license will result in the termination of the holder's activity.[246]

256. Besides the aforementioned situations, NRAE is entitled to withdraw an authorization or a license in any situation where an insolvency proceeding has been opened against the holder or at the holder's request.[247]

244. Article 26, para. 1 of Regulation concerning granting of authorizations and licenses, approved by Order 34/2013.
245. Exceptions from these rules are established by Art. 26, para. 4 of Regulation concerning granting of authorizations and licenses, approved by Order 34/2013.
246. Article 26, paras 2 and 3 of Regulation concerning granting of authorizations and licenses, approved by Order 34/2013.
247. In the latter case, if the activity is performed under a concession agreement, the withdrawal will take place only if the holder submits together with its request, the approval of the co-contractor. Article 27 of Regulation concerning granting of authorizations and licenses, approved by Order 34/2013.

Chapter 3. Rights and Obligations of License Holders in the Natural Gas Sector

§1. Common Rights and Obligations of License Holders in the Natural Gas Sector

257. All holders of licenses must comply with the following set of obligations: to maintain within the internal accounting separate records for each type of regulated or performed activity; to prepare, submit for audit and publish annual financial statements, according to 4th 78/660/CEE Directive; to maintain confidentiality of commercial information received from third parties; to not abuse commercially sensitive obligations obtained from third parties; to provide necessary data for third parties' access to the system in a clear, transparent and easily accessible manner; to hold all licenses and authorizations required by law; to make available at NRAE's request copies of all contracts of purchase/sale/supply and transit of natural gas.[248]

258. In addition to all the above, license holders bear the obligation to perform all activities with full observance of the license's or authorization's terms concerning safety, quality and continuous supply, energy efficiency, health and safety rules at work, environmental protection and the provisions of contracts concluded directly with customers. NRAE is also entitled to establish new obligations, specific to each activity related to natural gas sector, through regulations. However, these obligations must be transparent and non-discriminatory and should not impeach the liberalization of the natural gas market by creating barriers against entry of new operators on the market or by distorting competition. In order to ensure that none of these side-effects occur, NRAE notifies the Competition Council with respect to any obligation that might result in market distortion.[249]

§2. Rights and Obligations of the Grantees Concerning Goods Belonging to Third Parties

259. In order to perform the public service of transport or distribution of natural gas, the grantee enjoys the following rights with respect to goods belonging to third parties:

- To use these goods by law.
- To use these good as agreed by contracts or minutes concluded with the owners.
- To include the related costs concerning exploitation, maintenance, repair and modernization works in setting the tariff for the performed service.
- To take over the goods in full ownership, with the prior approval of their owner and payment of just compensation thereof.
- To develop the system.

248. For a complete list of obligations *see* Art. 122, para. 1 of Natural Gas Law.
249. Article 173, paras 1–4 of Natural Gas Law.

- To use the entire capacity of the goods.
- To ensure the access of new applicants (provided either the owner has given his acceptance or the applicant has agreed to pay compensation to the owner).
- To replace the goods where needed to expand or modernize the infrastructure or due to wear and damage.[250]

260. With respect to the grantees' obligations, the law specifically mentions that grantees must, on the one hand, finance, at expiration of the functioning period, the investments in replacing the goods and, on the other hand, must operate and maintain the goods at optimum level for safe functioning of the system. Additionally, in order to ensure safe and continuous supply with natural gas of customers grantees are not allowed to change the destination of goods belonging to third parties.[251]

§3. Right of Use and Easement Right: Other Rights

261. The grantees enjoy, for the entire duration of their concession: a right of use regarding the performance of the necessary rehabilitation and modernization of the objectives;[252] a right of usage for ensuring the functioning of the system at normal capacity by performing maintenance, repairs and interventions;[253] an easement right for installing networks, pipelines, lines or other related equipment and for accessing their location;[254] a right to obtain the restriction and cessation of any activities that might put in danger persons or goods;[255] and a right to access public utilities.[256] These rights cover land and any other public or private property goods, belonging to natural or legal persons as well as activities performed by natural or legal persons in the proximity of the capacities.[257]

262. The right to use and the easement right concerning public goods, are established by law and can be exercised without registration in the Land Registry or other

250. Article 108, para. 1 of Natural Gas Law.
251. Article 108, paras 2 and 3 of Natural Gas Law.
252. According to Art. 110 of Natural Gas Law the right to use for performance of works allows the grantee to store materials and equipment on private property land, to destroy or limit plantations or other appliances within the limits of the works performed, to extract materials or capture water, to install equipment and build offices or site dwellings or to stop or restrict the owner's activities.
253. According to Art. 111, para. 2 of Natural Gas Law the right to use for normal functioning of the system entitles the grantee to store and install materials and equipment or to destroy existing plantations and restrict the owner's activities within the territorial and time limits needed for performing the works.
254. According to Art. 112 of Natural Gas Law, under the easement right, the grantee is entitled to install networks, pipes, lines, pillars, and other equipment, or to access their location for interventions, maintenance, repairs.
255. According to Art. 114, para. 2 of Natural Gas Law, the operator of the distribution system is entitled to intervene at any moment over the natural gas distribution networks to perform urgent repair works, according to NRAE's regulations, without prior approvals or authorizations, in order to remedy those defects which endanger the safety of consumers.
256. According to Art. 115 of Natural Gas Law, the right to access public utilities must be exercised in good faith and without affecting the access of other persons to those utilities.
257. Article 109 of Natural Gas Law.

formalities for the entire duration of the natural gas objective or system. However, for the purposes of establishing fair compensation and reparation, the grantees conclude with the owners of the affected lands a master-contract whose content is established by Government Decision within thirty days from the latter's request. If during the works additional damages are caused to the owners or their neighbours, the grantees must pay for the coverage of those damages.[258]

263. However, in order to protect the natural gas objectives and systems, third parties are banned from carrying out any construction works, to store any materials within the *safety area* or intervene in any way on the pipelines, equipment or installations. In addition, third parties may only perform digging or other types of works within the *protection area* with the prior approval of the system operator.[259] Failure to comply with these obligations may result in criminal liability.[260]

258. Article 113 of Natural Gas Law. Damages are set on the basis of several legal criteria such as: the affected surface, the value of estimated crops and plantations as established by competent bodies and the market value of the affected goods. Where parties fail to reach an agreement, the amount of compensation is to be set by the court.
259. Article 190 of Natural Gas Law.
260. Article 193, para. 1 of Natural Gas Law.

Chapter 4. Production and Exploitation

264. In order to project, perform and exploit objectives or systems of natural gas sector, the legal persons must hold an authorization issued by NRAE. Same obligations apply in case of modification, modernization or expansion works regarding natural gas objectives. For the purpose of projection and performance, natural persons must have the status of authorized technicians, which is granted by NRAE on the basis of a specific regulation.[261]

265. The natural gas producer is the natural or legal person whose activity is production of natural gas, biogas or bio methane, or any other types of gas under the conditions of the law.[262]

266. The law established a series of obligations for the natural gas producer, such as: to hold the authorization for building new upstream pipelines and the license to operate them; to ensure the operation of upstream pipelines in a safe and efficient manner with full compliance of environmental protection rules; to ensure non-discriminatory access of third parties to upstream pipelines.[263]

267. Among the main rights of producers, the law mentions: the right to elaborate technical and commercial guidelines specific to their own activity and submit them to NRAE for approval, to commercialize the extracted gas; to stop the systems' functioning for the period required for performing maintenance and repair works, with prior notification of affected system dispatchers or final consumers, as the case may be; and to deny access or connection to third parties to the upstream pipelines under the conditions established by the law.[264] Denial of access must always be in writing and motivated[265] otherwise it constitutes misdemeanour and is sanctioned with a monetary fine.[266]

261. Article 121 of Natural Gas Law. Currently, the Regulation regarding the authorization of natural persons who perform activities in the natural gas sector dates from 10 Sept. 2014 and has been published in the Official Gazette of Romania, Part 1, No. 692/22.09.2014.
262. Article 123 of Natural Gas Law.
263. For a complete list of producers' obligations *see* Art. 124, para. 1 of Natural Gas Law.
264. Article 124, para. 2 of Natural Gas Law. Denial of access or connection to upstream pipelines is not arbitrary but must comply with the conditions set out in the law at para. 2, letter d), number 1–4, such as technical incompatibilities, difficulties that might prejudice the production, etc. Additional situations for denial are established in Art. 149 of the Law and refer to: insufficient capacity of the system; situations where the access to the system might impeach the performance of the public service or might cause serious financial or economic difficulties related to take or pay contracts entered by the license holder, or where the quality of natural gas does not comply to the legal requirements.
265. Article 149, para. 2 of Natural Gas Law.
266. Article 194, point 18 corroborate with Art. 195, para. 1, point 2, letter c) of Natural Gas Law.

Chapter 5. Transportation

268. The transport of natural gas is deemed to be a public service of national interest and it is realized by the transport and system operator, certified by the competent authority. By law, the national transport system of natural gas is the property of the state and any works conducted for its development, rehabilitation, modernization, exploitation and maintenance are works of public utility.[267] The terms and conditions of usage of the national transport system are set in the Network Code of the National Transport System.[268]

269. The operator of the transportation system is obliged to elaborate investment and development plans of the transport system for ten years, on the basis of the current state and predictions of future consumption and reserves of natural gas, including imports and exports. These plans must specify the financing and implementation methods and must be approved by NRAE.[269]

§1. The National Transport System: Certification of the System Operator

270. The operator of the National Transport System is organized and functions as an independent system operator and must be a legal person certified[270] by NRAE. The certification procedure of the transport and system operator is finalized with full observance of the provisions of Article 3 of CE Regulation 715/2009. NRAE issues a preliminary certification within 120 days from the submission of the operator's request, which is notified to the European Commission together with the related documentation. However, if no preliminary decision is rendered within the specified term, the certification is tacitly granted, and must also be notified to the European Commission together with the entire documentation attached. The designation of the system operator is approved by the European Commission, after the conclusion of the certification procedure.[271]

---

267. Article 125, paras 1–4 of Natural Gas Law.
268. Article 1 of the Network code for the national transport system of natural gas (hereinafter the Network Code), published in the Official Gazette of Romania, Part I, No. 171 bis/29.03.2013 and amended by NRAE President's Order No. 120/2013, Order No. 53/2014 and Order No. 88/2014.
269. Article 125, paras 6–8 of Natural Gas Law. According to para. 14, NRAE monitors all investment plans and submits an annual evaluation report of these planned investments concerning their compatibility with the community's plan of network development established by Art 8, para. 3 of CE Regulation 715/2009. In order to ensure compliance NRAE is entitled to recommend changes to the operators' plans.
270. Certification requirements are set out in Art. 128 of Natural Gas Law, while the procedure is detailed in Art. 126^2 of Natural Gas Law. For operators controlled by third countries, special rules concerning certification apply. These rules are mentioned in Art. 127 of Natural Gas Law and refer to NRAE's obligation to obtain the approval of the European Commission which must attest that the third country operator fulfils all legal conditions and does not put in danger the safe supply of natural gas to the European Union.
271. Article 126, paras 1–6 of Natural Gas Law.

§2. RIGHTS AND OBLIGATIONS OF THE SYSTEM OPERATOR

271. The system operator enjoys the right: to levy non-discriminatory tariffs for the performed

272. services,[272] to limit and/or to interrupt the service according to the regulatory provisions; to deny access of third parties to the transport system within the boundaries of the law; to interrupt or limit the transport of natural gas where safety measures so require; to interrupt the functioning of facilities for the period of time necessary to conduct repairs or maintenance works; to use freely, public property lands occupied by the national transport system facilities; to administer the balancing market; and to participate in the purchase of natural gas for the purpose of ensuring the physical balance of the system.[273]

273. According to Romanian law, the system operator also has to comply with a long list of obligations: to operate the system and ensure its residual physical balance[274] as well as the programming, dispatching and functioning of the system in full safety; to maintain, rehabilitate, modernize and develop the system safely and efficiently while also protecting the environment; to implement, maintain and develop a monitoring, command and data purchase system; to ensure non-discriminatory access or connection of third parties to the transport system;[275] to hold in underground deposits or to ensure the purchase of necessary quantities to ensure the physical balance of the transport system; to organize and administer the balancing market for natural gas.[276]

274. Additionally, the system operator must notify NRAE of any planned transaction which might require the re-evaluation of its compliance with the requirements set out in Article 128 of Natural Gas Law or with any other circumstances on the basis of which a person or several persons from one or more third countries would gain control either over the national transport system or over the system operator.[277]

275. The system operator performs also a set of activities for the purpose of informing interested parties with respect to: the management of measuring equipment and the measuring service of natural gas for the users of the system; the collection, centralization and distribution of data and information towards undertakings

272. The provisions covering the tariffs perceived by the system operator are contained in Arts 95–107 of the Network Code. There are several tariffs to which the system operator is entitled: transport tariffs, tariffs for exceeding the reserved capacity, tariffs for not reaching the reserved capacity, and tariffs for daily unbalance.

273. Article 130, para. 2 of Natural Gas Law.

274. The procedures governing the physical and commercial balancing of the national transport system are described in detail in Arts 82–90 of the Network Code.

275. Article 22 of the Network Code.

276. For a full list of system operator's obligations *see* Art. 130, para. 1, letters a)-u) of Natural Gas Law.

277. Article 129, para. 1 of Natural Gas Law.

which use the transport system for their commercial operations in order to handle any unbalances between the contracted gas quantities and those physically delivered; the elaboration of gas consumption forecasts for clients which do not benefit from daily measurements and the drawing up of reports and statistic data as required by law.[278]

276. Failure to comply with the legal obligations constitutes misdemeanour and is sanctioned with a fine according to the law.[279]

§3. Duties of the System User

277. Given the fact that the transport system operator grants capacity at the access/exit points in/from the system on a 'first come, first served' basis, system users reserve capacity[280] by entering a transport contract[281] with the transport system operator. The transport contract must be in accordance with the framework contract provided by the Network Code.[282] However, system users may not solicit more capacity than necessary for fulfilling their contractual obligations, storage obligations and own consumption.[283]

278. In order to facilitate the implementation of the transport contract, the system user is obliged to inform the transport system operator with respect to all deliveries and pickups of natural gas from the national system planned at all entry and exit points where the user has reserved capacity.[284] These information shall be contained in a Transport Schedule,[285] with full observance of all terms and procedures set in the Network Code. The transport schedule forms an integral part of the transport contract.[286]

§4. Duties of the Owner of the Transport System

279. The owner of the transport system must cooperate closely with the transport and system operator by providing the latter with all relevant information and by financing (or approving the financing by third parties or even the system operator) the investment decisions made by the independent operator and approved by the

278. Article 130, para. 5 of Natural Gas Law.
279. *See* the combined provisions of Arts 194 and 195 of Natural Gas Law.
280. The procedure for soliciting capacity is described in Arts 37–42 of the Network Code.
281. The conclusion of the transport contract is conditioned by financial and technical obligations, both detailed in Art. 27 of the Network Code. According to Art. 28, failure to fulfil the financial or the technical obligations throughout the existence of the contract may also constitute grounds for contract termination.
282. Article 25, para. 1 corroborated with Art. 26, para. 1 and Annex 1 to the Network Code.
283. Article 31 of the Network Code. Exceptions from these limitations are set in Art. 32.
284. Article 45, para. 1 of the Network Code.
285. The relevant provisions concerning the transport schedule are contained in Arts 46–48 of the Network Code. The model programme is described in Annex 5 to the Network Code.
286. Article 47 of the Network Code.

NRAE. It will also offer guarantees for facilitating investments of potential expansions of the system, except for investments agreed to be financed by third parties or the system operator. The owner is liable for the transport system's assets, however it will not be responsible for the activity of providing third parties with access to the transport system or for planning investments. Last, but not least, the owner of the system must maintain confidentiality of all commercially sensitive data obtained throughout its activities.[287]

287. Article 131 of Natural Gas Law.

Chapter 6. Distribution

280. The distribution of national gas is deemed to be a public service of general interest[288] and is granted by concession for one or more designated areas, namely administrative territorial units. The concession is exclusive for the granted areas.[289] The grantor together with NRAE, must ensure, that the concession of several adjacent administrative territorial units is done through a sole contract.[290] The distribution of natural gas is performed by the distribution operator, for all users of the distribution system. The access to the distribution system must be ensured in a non-discriminatory manner for all applicants which fulfil the legal requirements.[291] The distribution operator must be an independent entity,[292] like all other operators within the natural gas sector.

281. The granting authority is the relevant ministry,[293] which also elaborates the general framework governing the legal regime of concession contracts concerning the public service of natural gas distribution, the procedures for granting concessions as well as the content of the tender specifications. However, the Government must approve them by Decision before they are set into force.[294]

282. The granting process starts with a request received by the granting authority from an interested party or public body, under the conditions set out in the law.[295] The request must be accompanied by a feasibility study approved by the NRAE.[296] In case the request is successful, and a concession has been

288. Article 135 of Natural Gas Law.
289. Article 104, para. 1 of Natural Gas Law. According to para. 6 the concession may be expanded to other localities within the concession area.
290. Additionally, the relevant ministry and the competent authority must ensure that the supplying of the concession area is done through a single connection. Article 104, para. 5 of Natural Gas Law.
291. Article 134, paras 1 and 2 of Natural Gas Law.
292. In case the distribution operator is part of an undertaking vertically integrated, then the former must be independent from the latter at least with respect to its legal structure, organization and decision process regarding all activities unrelated to distribution. However, this rule is not to be constructed in the sense that there must be a separation of ownership over the assets which belong to the distributor and those of the undertaking. *See* Art. 135, para. 1 of Natural Gas Law.
293. Article 1, para. 1 of The Guidelines concerning the granting of the public service of distributing natural gas have been published in the Official Gazette of Romania, Part 1, No. 709/29.09.2014 (hereinafter Guidelines concerning distribution).
294. Article 104, paras 1^1 and 1^2 of Natural Gas Law.
295. The details concerning the documentation which needs to be attached to the applicant's request are set in Article 4 of the Guidelines concerning distribution. On the basis of the said documentation, the relevant ministry must prepare a study for substantiating the concession decision. This study also constitutes the basis for preparing the granting documentation. *See* Arts 5–7 and Annex A of the Guidelines concerning distribution.
296. Article 104, para. 2 of Natural Gas Law and Art. 3, paras 2 and 3 of the Guidelines concerning distribution. According to Art. 104, para. 3 of Natural Gas Law, the feasibility study must be drawn in accordance with NRAE's Methodology, by an undertaking authorized by NRAE in this regard; it must be approved by NRAE's decision and is part of the tender documentation. NRAE's approval is mandatory for the initiation of the process (para. 4). The Methodology concerning the drawing up of feasibility studies and the request of NRAE's approval in order to initiate the granting process of natural gas distribution has been issued by NRAE on 7 Jun. 2013 and has been published

granted,[297] the winner is expected to sign the concession agreement[298] and to solicit and obtain from NRAE all licenses and authorizations provided by the law in order to perform the activities.

283. The grantee of the public service of natural gas distribution must supervise the performance and completion of works undertaken through the concession agreement. The supervision can be done either through its own or through contracted personnel, which has been authorized thereof according to legal requirements.

284. The concession agreement institutes an automatic putting into delay, without the possibility of a grace period, for any failure to perform undertaken investments within the agreed timeframe. Therefore, the grantee owes delay damages, starting with the default date, equal to the amount of revenues calculated for the quantity of gas which should have been delivered.[299]

285. Failure to perform the agreed amount of works as established in the contract, lack of compliance with the agreed key terms of the contract (which mandatorily include the provisions concerning the payment of revenues and environmental protection), as well as the systematic violation of the validity conditions of the natural gas distribution licenses or of the legislation regarding the safe functioning of the objectives may also result in withdrawal of the concession rights together with the correlative licenses. The decision to withdraw the concession may be contested in administrative courts, within sixty days from its issuance.[300]

286. In the case where concession contracts are terminated for any reason, the grantor or any other grantee is entitled to take over all assets related to the public service of natural gas distribution, which are property of the grantee, with his approval, for a payment equal to the regulated amount which was not amortized yet, established by NRAE.[301]

in the Official Gazette of Romania, Part 1, No. 388/10.06.2013. The Methodology applies to undertakings authorized by NRAE to draw feasibility studies (Art. 2) and sets out the steps to be followed to obtain the feasibility study and the approval of NRAE in order to initiate the procedure for the granting of the concession (Art. 1). The Methodology establishes the minimum content of the feasibility study (Art. 5) and the procedure for obtaining NRAE's approval (Arts 6–8) as well as the format of annexes and requests which need to be submitted (Art. 10).

297. Concessions are granted as a result of a public tender offer. The procedure concerning the public tender is described in detail in Art.s 8–15 and Annex B of the Guidelines concerning distribution.
298. According to Art. 16, para. 1 of the Guidelines concerning distribution the concession agreement is concluded on the basis of the provisions of the framework contract contained in Annex C to the Guidelines.
299. Articles 18 and 19 of the Guidelines concerning distribution.
300. Article 106 of Natural Gas Law.
301. Articles 107 and 126, para. 8 of Natural Gas Law.

§1. Right and Obligations of the Distribution Operator

287. The distribution operator enjoys similar rights to those of the transport and system operator. Therefore, the distribution operator is entitled to conduct commercial activities related to the distribution service; to cash in the tariffs for the services performed; to limit or stop the distribution service or the functioning of the equipment and the natural gas supply of the clients for the time necessary for conducting repairs and maintenance works or in case of force majeure, with prior notice to the dispatchers of the affected systems; to use, free of charge, the public property lands occupied by the distribution system's objectives; to stop the supply of natural gas where there is a danger for explosion; to decline connection to the distribution system within the boundaries established by the law[302] or to elaborate technical and commercial guidelines for its activities and submit them for approval to NRAE.

288. The law has established a non-exhaustive list of obligations for the distribution operator. Among them, the law states that the operator must operate, maintain, repair, modernize and develop the distribution system while complying with safety, economic efficiency and environmental protection measures; to ensure the odorizing of the natural gas according to NRAE's regulations, on the basis of the service contract concluded with the upstream operator; to interconnect with other systems; to ensure non-discriminatory access of third parties to the distribution system; to avoid cross-subsidizing between final consumer categories concerning allocation of costs; or to ensure safe distribution of natural gas.[303]

289. In addition to all the above, the distribution operator must maintain confidentiality over the commercially sensitive information obtained throughout its activity and it is banned from discriminatory disclosure of information regarding its own activities, which would be commercially advantageous.[304] Failure to comply with the obligation to maintain confidentiality constitutes misdemeanour and is sanctioned with a monetary fine.[305]

§2. Closed Distribution System

290. By closed distribution system, the law understands that system through which natural gas is distributed in an industrial, commercial or common services area, geographically limited, and which does not supply residential consumers, if, for technical reasons, the activities performed by the users of the closed system are

302. The situations in which the operator is entitled to decline a request for connection to the distribution system are set out in Art. 150 of Natural Gas Law and they refer to: the insufficient capacity of the system; lack of components for the systems which request connection; default in payment of connection tariffs.
303. For a complete list of duties *see* Art. 138, para. 1 of Natural Gas Law.
304. Article 137 of Natural Gas Law.
305. *See* combined provisions of Art. 194, point 40 and Art. 195, para. 1, point 2, letter c) of Natural Gas Law.

integrated or, if the said system ensures natural gas mainly to the owner of the system, the system operator or an affiliated undertaking.[306]

291. To set up or operate a closed distribution system one requires a set-up authorization or, respectively, a license issued by NRAE. The tariffs for the distribution service through a closed distribution system together with the methodologies for establishing the said tariffs are also to be approved by NRAE, at the operator's request.[307]

306. Article 139, para. 1 of Natural Gas Law.
307. Article 139, paras 3 and 5 of Natural Gas Law.

Chapter 7. The Natural Gas Market

292. According to the law, the natural gas market is the organizational framework where natural gas and ancillary services are traded on.[308] It comprises the regulated market and competitive (free) market, while transactions can be either wholesale or retail.[309] Romanian law mentions also the centralized natural gas market[310] and the natural gas balancing market. The first one is the framework for performing natural gas transactions between different undertakings, intermediated by an operator of the natural gas market, on the basis of its own specific rules[311] approved by the competent authority. The second one is the framework for concluding natural gas transactions between different undertakings, this time intermediated by an operator of the transport system, on the basis of specific rules approved by the competent authority.[312]

293. According to the data provided by NRAE,[313] the current structure of the Romanian natural gas market is the following:

- One operator of the National Transportation System: SNTGN Transgaz S.A. Medias.
- Six producers: OMV Petrom, Romgaz, Amromco, Toreador, Wintershall Medias, Aurelian Oil & Gas.
- Three operators for underground storage: Romgaz, Amgaz and Depomures.
- Thirty-four distribution and supply companies for captive consumers, the largest being Distrigaz Sud and E.ON Gaz Romania.
- Seventy-six suppliers for the wholesale market.

§1. The Functioning of the Regulated Natural Gas Market

294. The regulated market had as primary goals: the supply of natural gas, at regulated prices and on the basis of master agreements, until the end of 2014, to non-household customers,[314] unless there is a significant difference in the trading

308. Article 100, Point 71 of Natural Gas Law.
309. Article 174, para. 1 of Natural Gas Law.
310. The centralized natural gas market is governed by NRAE's General Rules of 12 Jul. 2013, published in the Official Gazette of Romania, Part I, No. 436/17.07.2013 and amended by NRAE's President's Orders 66/2014 and 6/2015 (hereinafter referred to as General Rules for CNGM).
311. According to Art. 4 of General Rules for CNGM the specific rules should contain at a minimum: the definition of the centralized market's purpose and scope; the description of the traded products; the framework for the organization and conduct of trading sessions; the applicable legislation and the information needed to be published in order to ensure transparency on the centralized market.
312. Article 100, Points 72–73 of Natural Gas Law.
313. *See* NRAE's website: http://www.anre.ro/ro/gaze-naturale/informatii-de-interes-public/piata-gazelor-naturale, last visited 23 Mar. 2018. The data is available in Romanian only.
314. NRAE's President's Order No. 107/2014 establishing a set of measures regarding natural gas supply to non-household customers in the view of eliminating final regulated prices, published in the Official Gazette of Romania, Part I, No. 780/27.10.2014, addressed the relationship between gas suppliers and their non-household customers which did not exercise their eligibility right and did

price of internal production and the European price for imports, which could affect the stability of the market, in which case the term is prolonged until the end of 2015; the supply for natural gas, at regulated prices and on the basis of framework contracts until mid-2021 to residential customers;[315] and the final stage supply.[316]

295. Other activities ancillary to the regulated market include: activities with a natural monopoly character, activities related to LNG terminals' operations, natural gas transport and storage in underground deposits or in pipelines, natural gas or biogas and bio-methane distribution. All these activities are performed at regulated tariffs established by the NRAE.[317]

296. The competent authority continuously monitors the effects of the regulated market over the competitive market and takes measures to ensure that there are no distortions of fair competition and for the increase of transparency with regard to commercial transactions.[318]

§2. The Functioning of the Competitive Market

297. Unlike the regulated market where prices are established by NRAE, on the competitive market there are either wholesale, or retail commercial transactions, and the prices are established on the basis of supply and demand principle. The wholesale competitive market is grounded on: bilateral agreements between undertakings in the natural gas sector; transactions on centralized markets, administrated by the natural gas market operator or the balancing market operator, as the case may be; or

not opt in for a negotiated supply regime. For the cases where non-household customers failed to conclude a new sale-purchase agreement with the suppliers, Art. 3 of Order No. 107/2014 established a tacit extension of the natural gas sell-purchase agreements already in force, with the sole exception of the provisions concerning the regulated price.

315. According to Art. 2 of Methodology for establishing the regulated unitary incomes and the total unitary incomes related to the regulated supply and for approving the regulated prices for the natural gas sector in 2015, published in the Official Gazette of Romania, Part I, No. 121/17.02.2015 (hereinafter Methodology for regulated prices), the regulated supply of natural gas represents the commercial activity of selling natural gas to residential customers who have not exercised their eligibility right and covers all the activities of the supply license holder concerning: the purchase of natural gas, selling natural gas to residential customers on the basis of framework contracts and at a regulated price and the relationship with residential customers determined by the conclusion and performance of regulated supply contracts. For those residential customers who did not exercise their eligibility right, the natural gas supply shall be performed at prices regulated by NRAE according to the aforementioned Methodology.

316. Article 175, para. 1 corroborated with Art. 179, para. 2, letters a)-c) of Natural Gas Law.

317. Article 179, paras 1, 2, letters e)-k) and 4 of Natural Gas Law. Prices are established on the basis of methodologies elaborated and approved by Order of NRAE's President and published in the Official Gazette of Romania, with prior information and consultation of all stakeholders. *See* the Methodology for regulated prices.

318. Article 175, para. 4 of Natural Gas Law. The criteria for evaluating the functioning and the effects of the regulated market are detailed in para. 5.

other types of transactions or agreements. On the retail competitive market supplier sell natural gas to final consumers based on contracts with negotiated prices or standard deals.

§3. Prices and Tariffs

298. The Romanian system of prices and tariffs aims to achieve: fair and full competition on the natural gas market; diversification of natural gas sources and the increase of natural gas supply safety; prudent recovery of costs without affecting the profitability of the investments;[319] stimulation of development of production, storage, distribution and supply of natural gas and LNG; energy saving for final consumers and increase of quality of natural gas and ancillary services.[320]

299. In order to achieve the aforementioned goals, the Romanian system of prices and tariffs is based on a series of principles such as: non-discriminatory treatment, transparent and objective criteria and prior public consultation of stakeholders. All prices and tariffs are set in full accordance with guidelines published by NRAE and are meant to stimulate efficient usage of natural gas, to promote competition and protect the interests of consumers, and prevent profiteering. Therefore, recovering costs related to services performed on behalf of a certain category of final consumers by inflating costs for other categories of final consumers as well as cross-subsidies between regulated activities or between regulated and unregulated activities is strictly banned.[321]

300. In case there are major disparities between supply and demand or obvious malfunctions of the natural gas market, the Government may, based on NRAE's proposal and with consultation of the Competition Council, limit the excessive increase of prices and tariffs or freeze them for a determined period of time,[322] not longer than six months, which may be prolonged for three additional months as long as the circumstances that determined the Government's intervention persist.

319. According to Art. 173, para. 5 of Natural Gas Law, all prudent investments made by undertakings in order to comply with their public service obligations are deemed justified costs and may be recovered through prices and tariffs, in accordance with NRAE's regulations.
320. Article 178, para. 1 of Natural Gas Law.
321. Article 178, paras 2 and 3 of Natural Gas Law.
322. Article 180, para. 1 of Natural Gas Law established two methods of intervention: either by setting a maximum threshold of the price or by limiting the amount of income resulting from the regulated activity.

Chapter 8. Storage, Supply and Consumption

301. The storage of natural gas is performed by the storage operator for the purposes of: securing natural gas supply to final consumers; harmonizing the variations of seasonal, daily and hourly consumption with natural gas sources available; ensuring physical balance of the national transport system and performing other commercial activities.[323]

302. As in the case of all other system or distribution operators, the storage operator must be an independent undertaking. Hence, in case the storage operator is part of an undertaking vertically integrated, then their operations, organization and decision making process need to be unbundled.[324]

§1. Rights and Duties of the Storage Operator

303. The storage operator is entitled to cash in the tariffs for the service of underground storage of natural gas and to cut off or limit the storage service according to the applicable regulations. It is also entitled to elaborate technical and commercial guidelines for its activities and to submit them for approval to NRAE. Similar to all other operators, the storage operator may stop the functioning of the equipment for the time needed for repairs or maintenance works with prior notice of dispatchers of affected systems or clients, as the case may be. Last but not least, it is entitled to refuse, for cause, the access of third parties to storage facilities.[325]

304. The main obligations required by the law are also similar to those of other operators. The storage operator must: operate, maintain, rehabilitate, and modernize the surface technological facilities corresponding to the storage deposits; ensure access of third parties to storage deposits on the basis of objective, transparent and non-discriminatory criteria; publish a list of storage facilities which are offered for access to third parties; furnish information to the users of the storage system; elaborate and submit to NRAE investment plans for five years regarding the systems they operate and ensure appropriate means for complying with the demand of their public service. In addition, the storage operator bears an obligation to maintain confidentiality and not disclose any commercially sensitive data obtained throughout the course of its activity.[326]

305. According to the law, namely Article 8 of the Methodology concerning the annual minimum stock of stored gas,[327] the holders of supply licenses and the holder of the license to operate the national transport system of natural gas, must maintain

323. Article 140 of Natural Gas Law.
324. Article 141 of Natural Gas Law.
325. Article 142, para. 2 of Natural Gas Law.
326. Article 142, paras 1 and 3 of Natural Gas Law.
327. The Methodology concerning the annual minimum stock of stored gas (hereinafter Minimum Stock Methodology) was approved by NRAE President's Order No. 35/2016, published in the Official Gazette of Romania, Part I, No. 552/21.07.2016.

a certain annual stock of natural gas in their deposits, [328] which is established by Order of NRAE's president.[329] The purpose is to ensure the natural gas supply of final clients and, respectively, the physical balance of the national transport system.[330] The minimum total stock is set so that the natural gas stored is equivalent to at least 25% of the total consumption of natural gas extraction recorded during the previous cycle storage or during 1 October to 31 March.[331] Failure to comply with the minimum levels of stock set by NRAE constitutes misdemeanour and is sanctioned according to the provisions of Natural Gas Law.[332]

§2. Rights and Duties of the Natural Gas Supplier

306. Although mentioned in the law, the actual rights of the gas supplier are of contractual nature. The gas supplier is entitled to cash in the price for the supplied natural gas (together with any delay damages); to cash in the price for the natural gas supplied under a last stage regime, according to NRAE's regulations; to check the meters whenever deemed necessary with prior notification of the final consumer;[333] and to limit or cease the supply with natural gas to customers, under the terms specifically mentioned in the contracts.[334] Additionally, the supplier is entitled to require the transport or system operator to stop the supply to the final consumer where the supplier has observed actions meant to affect the indications of measuring instruments or to steal natural gas and the final consumer has declined the supplier's request to provide a financial guarantee.[335]

307. Unlike its rights, the main duties of the gas supplier are specifically established by law. However, the list is not exhaustive and parties may establish additional duties by means of contract. Among the main obligations of the supplier, the law mentions: concluding natural gas purchase contracts in order to ensure the supply of its customers as well as the more recent obligation that the gas for supplying residential clients must be purchased in such conditions that it minimizes the cost

328. Article 2 of the Minimum Stock Methodology.
329. Currently, there are two Methodologies in place for determining the annual minimum stock, one for holders of licenses for transport system operators (Order No. 36/19.07.2016, published in the Official Gazette of Romania No. 551/21.07.2016 (hereinafter Order No. 36/2016)) and one for holders of supply licences (Order 35/19.07.2016, published in the Official Gazette of Romania No. 552/21.07.2016 (hereinafter Order 35/2016)).
330. Article 1 corroborated with Art. 3of the Minimum Stock Methodology.
331. Article 6, para. 1 of Natural Gas Law.
332. Article 15 of Natural Gas Law.
333. Part III 'On the contract for regulated supply of natural gas to residential customers', Point 2 'Rights and Obligations of the consumer and supplier of natural gas', para. 2.3 'Rights of the supplier', letter b) at http://www.anre.ro/ro/gaze-naturale/consumatori/consumatori-gn-racordare-contractare-raspunsuri-cadru, last visited on 23 Mar. 2018. Text available in Romanian only.
334. Article 143, para. 2 of Natural Gas Law.
335. Article 145, para. 3 of Natural Gas Law.

of allocated resources, on the basis of transparent purchase process, and by ensuring equal and non-discriminatory treatment among the offerors;[336] paying for the purchased gas, according to the concluded contracts; performing the supply activities on the basis of contracts concluded within the boundaries of NRAE's regulations; complying with performance standards concerning natural gas supply; allowing customers to freely switch to another gas supplier within twenty-one days from the date of the request; informing customers of any changes of contractual provisions or tariffs; providing customers with several methods of payment; refraining from unfair or deceptive commercial practices; storing natural gas at necessary levels to ensure continuity in supply to its customers.[337]

§3. Duties of the Supplier of Last Resort

308. For the supplier of last resort of natural gas, the Romanian law has expressly established two main obligations. On the one hand, the supplier of last resort must ensure natural gas supply to final consumers, according to NRAE's regulations, at prices regulated by NRAE. On the other hand, the supplier of last resort must supply natural gas to final consumers whose supplier is on the verge of losing its supply license or any other similar situation in which final consumers do not have any other source for natural gas supply.[338] These obligations are in fact protection measures of final consumers with regard to the need for natural gas and are only ancillary to the services of suppliers of last resort.

309. Suppliers of last resort bear the obligation to inform final clients with respect to the commercial conditions governing the supply of natural gas.[339] This obligation covers both the pre-contractual phase and the contractual one and must ensure in a transparent manner, clear, complete and legible data concerning the contract. However, the methods which the suppliers can make use of differ from one

336. Article 142, para 1, letter a[1]) of Natural Gas Law, amended by GEO No. 64/2016 for amendment and completion of Electrcitity and Natural Gas Law No. 123/2012, published in the Official Gazette of Romania, Part I, No. 801/11.10.2016 (hereinafter GEO No. 64/20126.
337. For a complete list of duties *see* Art. 143, para. 1.
338. Article 144 of Natural Gas Law.
339. Article 2 of NRAE President's Order 106/2014 concerning the methods of natural gas suppliers to inform final consumers with respect to commercial conditions regarding the supply of natural gas (hereinafter Order 106/2014), published in the Official Gazette of Romania, Part I, No. 781/ 27.10.2014.

stage to the other. In the pre-contractual phase, suppliers may use their own methods[340] or the interactive web application set up by NRAE,[341] while in the contractual phase, suppliers may use the contracts,[342] the invoices and the attached documents.[343]

§4. Rights and Duties of Final Consumers of Natural Gas

310. According to Article 100, point 25, final consumers are those who purchase natural gas for their personal consumption. Based on this definition, it appears that this broad category will include both residential customer[344] as well as industrial customers,[345] leaving aside wholesale customers,[346] who purchase natural gas for the purpose of reselling it inside or outside the system in which they are established.

311. Final consumers enjoy a wide range of rights such as: access to natural gas objectives/systems and the right to be informed when connected to them with respect to their right to be supplied with a specific quantity of natural gas, at reasonable prices; right to choose and negotiate a contract directly with their supplier (eligibility right);[347] to conclude a contract with the supplier;[348] to receive a free copy of the contract; to solicit the supplier or system operator the modification or amendment of the contract when new circumstances so require; to be notified of any intention to amend the contract's terms; to denounce the contract where the new terms are not acceptable; to have several methods of payment of their disposal; to freely switch their supplier;[349] to receive a liquidation invoice within forty-two days

340. Suppliers' own methods are detailed in Arts 3 and 4 of Order 106/2014. According to para. 2 of Art. 3, the supplier must provide for free a printed or electronic copy of the contract before the signing, if so requested by the final consumer. In addition, in order to allow the consumer to make an informed decision, all suppliers must publish on the internet or make available at their contact points standard offers. The information to be contained in standard offers is detailed in Art. 4, para. 6.
341. The characteristics of NRAE's web application are detailed in Art. 5 of Order 106/2014.
342. The minimum content of contracts is stated in Art. 6, para. 1 of Order 106/2014.
343. The data which the suppliers must make available through invoices and the attached documents are mentioned Art. 7 of Order 106/2014 and cover: the component elements of the final price; the method of determination of the final price; information concerning the right of the final consumer to change its supplier; and information concerning the disputes.
344. Article 100, point 23 of Natural Gas Law.
345. Article 100, point 26 of Natural Gas Law.
346. Article 100, point 22 of Natural Gas Law.
347. Article 174, para. 6 of Natural Gas Law.
348. According to Art. 145, para. 4, letter b) the contract between final consumer and the supplier must include at a minimum the following information: the identity and the address of the supplier, the services offered and their quality, the starting term of supply, the maintenance services offered, the means to obtain information on the applicable prices and tariffs, the contract's duration, conditions for extension, suspension, interruption, whether there is a right of unilateral termination, compensations, and the initiation procedure of dispute resolution.
349. The right to change the natural gas supplier belongs to both household and non-residential customers, although there are different methodologies applicable for each category. For residential customers, the applicable methodology was approved by NRAE President's Order No. 47/2007,

from changing their supplier; to benefit from transparent, simple and low cost dispute resolution proceedings; and to solicit a cease of supply and repairs where technical malfunctions affect the safe supply of gas.[350]

312. Given their status final consumers are precluded from selling natural gas. They also have the obligation to pay their consumption bills within the terms and conditions established in the contract. Failure to pay the bills within the due dates entitles the supplier to seek damages.[351] Additionally, where they have exercised their eligibility right for choosing their supplier, final consumers are precluded from returning to regulated supply.[352] However, notwithstanding whether they have exercised their eligibility right or still belong to the regulated market, residential customers enjoyed the same treatment with respect to the quality and the price of natural gas, until 31 of March 2017.[353]

313. Final consumers also have the obligation to solicit and obtain all legal approvals prior to any modification of the natural gas equipment or installation and to inform at once the supplier when they notice any malfunction of the measuring equipment or the natural gas installation.[354]

§5. Evolution of Production and Consumption

314. In Romania natural gas is produced almost exclusively (97.3%) by two companies: Romgaz and OMV Petrom, while the remaining production is represented by other companies. Almost 62.5% of the total gas coming from domestic production is extracted from Mures County. As for imports, Romania's main source is Russian Gazprom, which provides about 20% of the natural gas needed for consumption. Since the prices imposed by Gazprom are extremely disadvantageous, Romania sought to expand exploration works, in order to achieve energy independence. Currently fifty-nine oil and gas concession agreements are in place.[355] To the great disappointment of the authorities (but for the great satisfaction of the local

published in the Official Gazette of Romania, Part I, No. 796/2007 (hereinafter Order 47/2007) while for non-residential customers, the methodology was published in the Official Gazette of Romania, Part I, No. 385/21.05.2008. However, the stages of the change are similar for both categories: finding a new natural gas supplier; concluding a contract with the new supplier and rescinding the one with the current supplier; payment of all amounts due to the current supplier; and determination of the meter's index with the new supplier.

350. For a complete list of rights *see* Art. 145, para. 4 of Natural Gas Law.
351. Article 145, paras 1, 2 and 5 of Natural Gas Law.
352. Article 174, para. 7 of Natural Gas Law.
353. Article 177, para. 3^3 of Natural Gas Law, as modified by GEO No. 64/2016.
354. http://www.anre.ro/ro/gaze-naturale/consumatori/drepturile-si-obligatiile-consumatorilor, last visited on 23 Mar. 2018. Text available only in Romanian.
355. Alexandra Pele – Harta intereselor energetice ale americanilor si rusilor in Romania, Gandul.info, 18 Oct. 2013, available online in Romanian only at: http://www.gandul.info/financiar/harta-intereselor-energetice-ale-americanilor-si-rusilor-in-romania-11528530, last visited 23 Mar. 2018.

people who opposed the prospections) the results of Chevron's exploration well for shale gas proved to be not commercially viable and the company abandoned further works.[356]

315. According to the latest public report of NRAE, in 2016, Romanian internal production of natural gas that was used for consumption, represented 87.5% of the total amount of sources, with Romgaz and OMV Petrom providing 93.5% of this percentage. Imports amounted for 12.5%. Year 2016 represented a significant increase of consumption from imported sources, due to a decrease of quantities produced internally, but also due to a decrease of prices in imported quantities in the summer of 2016.[357]

316. Before 2012 it was expected that the balance between imports and domestic production will start to reverse due to exhaustion of domestic resources after 2015. However, recent discoveries in the Black Sea increased again the hopes for energy independence, at least with respect to natural gas necessities,[358] although these deposits are still far away from being exploited. Recent information tends to point out that the gas reserves will be sold on external markets, provided that the necessary infrastructure will be finished on time.[359] In January 2015, Romania imported only 2% of its consumption necessity,[360] but this was only the result of a set of factors, the most important being the winding up of some of the large natural gas industrial consumers. One year later, the imports reached 12.5%.[361] Thus, Romania still faces the prospect of natural gas dependency, which will mean dependency on Russia, since Gazprom is the sole source of gas imports to Romania. The Government's attempts to diversify sources of gas imports by connecting with other major gas exporters, such as Azerbaijan or Kazakhstan, have, so far, been unsuccessful.[362]

---

356. Everett Rosenfeld, *European Shale Dream is Dying Before it Started*, 23 Feb. 2015, CNBC, available online at http://www.cnbc.com/id/102446971, last visited 23 Mar. 2018.
357. NRAE: Raport annual privind activitatea Autoritatii Nationale de Reglementare in Domeniul Energiei – 2016, pp. 221–223, available online, in Romanian only, at: http://www.anre.ro/download.php?f=hKh%2Fgg%3D%3D&t=vdeyut7dlcecrLbbvbY%3D, last visited 23 Mar. 2018.
358. Romulus Cristea – Descoperire istorica de gaz in Marea Neagra. Ce are de castigat Romania, 22 Feb. 2012, Romania libera, available online, in Romanian only, at http://www.romanialibera.ro/actualitate/eveniment/descoperire-istorica-de-gaz-in-marea-neagra--ce-are-de-castigat-romania-254582, last visited 23 Mar. 2018.
359. Andrei Visan – Gazele romanesti din Marea Neagra vor merge afara, 21 Feb. 2018, http://evz.ro/gazele-romanesti-marea-neagra.html, last visited 23 Mar. 2018.
360. Adrian Mosoianu – Romania va importa luna viitoare, ianuarie 2015, doar 2% din necesarul sau de consum de gaze naturale, Energy Report, 23 Dec. 2014, available online, in Romanian only, at: http://www.energyreport.ro/index.php/2013-petrol-si-gaze/2013-stiri-petrol-si-gaze/2013-rafinare-si-marketing/3401-romania-va-importa-luna-viitoare-ianuarie-2015-doar-2-din-necesarul-sau-de-consum-de-gaze-naturale, last visited 23 Mar. 2018.
361. *See supra* fn. 359.
362. Aurel Dragan – Dupa un secol de exploatare, mai avem gaze doar pentru 10 ani, Capital.ro, 30 Nov. 2013, available online, in Romanian only, at: http://www.capital.ro/189384.html, last visited 23 Mar. 2018.

Part III. Regulatory Framework Concerning Petroleum

317. Romania has a long-standing tradition concerning oil industry, going over 150 years back. In 1857, Romania became the first country in the world to officially record its oil production. In 1861, the first well drilled in Romania reached 150 m depth. By 1910, there were already foreign capital companies established such as Star Romanian, Romanian–American, Vega, Concordia or Astra Romanian. Romanian oil played an important part in both World Wars.[363] Since Romania was the largest source of oil for Europe (excluding USSR)[364] during WW1, the oil fields near Ploiesti were blown up after the invasion of German–Austrian troops,[365] while in WW2, the same fields have fuelled the needs of the German Army,[366] which led to heavy bombardments of the area by the Allied forces.[367]

363. *See* online data provided by NAMR on their website: http://www.namr.ro/petroleum-law/general-presentation-oil/. On the control of foreign capital companies over the incipient Romanian oil industry *see*: Daniel Yergin, *The Prize. The Epic Quest for Oil, Money and Power* 132 (Simon & Schuster 1991).
364. Yergin, p. 334.
365. *Ibid.* at 180–181.
366. *Ibid.* at 334 and 370.
367. *Ibid.* at 347.

Chapter 1. General Introduction

318. After the war, following the nationalization of mineral resources, the development and exploitation of petroleum reserves were undertaken by the state. In 1975, Romania inaugurated its first marine drilling platform located on the continental shelf in the Black Sea. The peak oil production of the country, 14.6 million tons, was achieved in 1976.[368]

319. Following the anti-communist revolution of 1989, in 1990, several state-controlled companies were established to act in the field: SC Oil Terminal Constanta (handling crude oil, liquid petroleum and petrochemical products and other goods, raw materials for import, export and transit) and Conpet SA Ploiesti (concerning the operation of the national system of pipelines to transport crude oil from the mining fields and the refining Oil Terminal). In the same year, the former Petroleum Ministry was abolished and transformed into a Department of the Ministry of National Economy. The National Agency for Mineral Resources (NAMR) was established in 1994 and in 1995 the first Petroleum Law was adopted. However, it referred exclusively to hydrocarbons: oil, gas and condensate.[369]

320. In 1997 the first national oil company was established, Petrom SA, which was to be privatized in 2004 as a pre-condition for the EU accession of the country. The majority pack of 51% was acquired by OMV Group from Austria. In 2004, another Petroleum Law was adopted with the purpose of aligning Romania to international standards.

321. The current general framework concerning petrol is the Petroleum Law No. 238/2004 (revised and amended) (hereinafter 'the Petroleum Law').[370] It is supplemented by Methodological Norms (hereinafter 'the Rules'), adopted through Government Decision No. 2075/2004[371] and by Orders, Decisions or Procedures of

368. *See* online data provided by NAMR on their website: http://www.namr.ro/petroleum-law/general-presentation-oil/.
369. *See* online data provided by NAMR on their website: http://www.namr.ro/petroleum-law/general-presentation-oil/.
370. Petroleum Law No. 238/2004 published in the Official Gazette of Romania, Part I, No. 535/15.06.2004. It repealed the previous Petroleum Law No. 134/1995, published in the Official Gazette of Romania, Part I, No. 301/29.12.1995 and any other contrary legal provisions (*see* Art. 65 of Petroleum Law No. 238/2004).
371. The Methodological Rules for the Application of Petroleum Law 238/2004 were adopted through Government Decision No. 2075/24.11.2004, published in the Official Gazette of Romania No. 1170/10.12.2004. It repealed Art. 1 and the Appendix to Government Decision No. 1265/1996 concerning the approval of Methodological Rules for the Application of Petroleum Law No. 134/1995 as well as the additions and amendments to Government Decision No. 1043/1995 concerning several measures for co-option of foreign partners in oil and gas exploitation in Romania, published in the Official Gazette of Romania, Part I, No. 320/2.12.1996 (*see* Art. 2 of Government Decision No. 2075/24.11.2004).

NAMR.[372] Additional legal provisions are to be found in the Romanian Constitution[373] and the NCC.[374] As Romania is also a Member State of the European Union from 2007, the EU legislation concerning oil is also to be taken into consideration. However, the focus of this part will be mainly on Romanian national regulatory framework.

372. *See* for example: Order No. 122/14.06.2006 concerning the approval of the Methodology for certification of technical competence of businesses that prepare documentation and/or geological research works, works of exploitation of petroleum and mineral resources, and of expertise, as well as individuals who prepare documentation and/or executing works of geological research and expertise (published in the Official Gazette of Romania, Part I, No. 551/27.06.2006); Order No. 53/21.03.2008 for the approval of Instructions concerning criteria, methodology and procedure for establishing regulated tariffs for the transport of crude oil, gasoline, condensate and liquid ethane through the National Transmission System (published in the Official Gazette of Romania No. 267/4.04.2008); Order 194/3.10.2006 for the approval of Instructions concerning the elaboration of the annual petroleum exploitation (published in the Official Gazette of Romania No. 854/18.10.2006); Decision No. 1137/04.05.2010 concerning the approval of the Evaluation procedure of the conditions necessary for the conclusion of contracts for performing interruptible services concerning transportation of natural gas through the National Transmission System (available online at: http://www.namr.ro/wp-content/uploads/2013/02/Decizia-1137-din-mai-20101.pdf) or Operational Procedure for granting prospecting licenses (2nd Revision) issued on 7 Jun. 2013 (available on line at: http://www.namr.ro/wp-content/uploads/2014/02/procedPP2.pdf).
373. The Constitution of 2003 was published in the Official Gazette of Romania No. 767/21.10.2003 and established the general regime concerning ownership of mineral resources, including petroleum (*see* Art. 136, paras 2–5).
374. The New Civil Code of Romania was adopted by Law No. 287/2009 and published in the Official Gazette of Romania No. 505/15.07.2011. It contains general provisions concerning public property (*see* Art. 554) and the rights deriving from it – the right to administer or to lease by concession (Art. 866–873).

Chapter 2. Production

322. As stated in the Constitution and the Petroleum Law,[375] the petroleum resources[376] situated in the subsoil and the Romanian continental shelf in the Black Sea constitute *exclusive public property of the Romanian state*. Strictly based on the Constitutional approach it may seem that the ownership regime applicable in Romania is 'state ownership', according to which hydrocarbon reserves are owned by the state and form part of its estate. However, under such a regime, the principle of state ownership would result in state monopoly, with companies acting as mere contractors, which is not the case anymore. Judging by the Petroleum Law's provisions, the fact that the state grants mineral rights by way of concessions to the company it chooses, based on a process involving competitive bidding, it would seem that the ownership regime has features of 'state discretion'. However, according to this regime petroleum reserves are not owned until they are discovered, which is not the case in Romania. Hence, it is safe to claim, based on the above, that *the applicable ownership regime in Romania is a hybrid one*, combining principles of both state discretion and state ownership, which brings Romania in line with the majority of the countries.[377]

323. After the privatization of the National Oil Company – Petrom SA, in 2004, the state relinquished its possibility of exploring and producing oil by direct action. Currently, *only indirect action is available*, for the state, as owner of mineral resources decides who will conduct petroleum operations,[378] subject to the relevant national legal framework and the applicable contractual regime.

324. The contractual regime adopted by the Romanian Petroleum Law is referred to as a 'concession' (or licensing),[379] under the form of a 'petroleum agreement',[380] concluded between the Romanian state, represented by the competent

375. Article 136 of the Constitution and Art. 1(1) of the Petroleum Law.
376. Petroleum Law specifically refers to petroleum and natural gas, each defined in Art. 1(2) respectively (3).
377. Nadine Bret-Rouzaut & Jean-Pierre Favennec, *Oil and Gas Exploration and Production – Reserves, Costs, Contracts* 172 (3d ed., Editions Technip 2011).
378. According to Art. 2, point 23 of Petroleum Law, petroleum operations are defined as 'all activities of exploration, development, exploitation and abandonment of an oil reservoir, underground storage, transportation and transit of oil through pipelines, and operation of oil terminals'.
379. According to Art. 2, point 7 of Petroleum Law, a petroleum concession is 'the legal operation by which the Romanian state, represented by the competent authority, as grantor, transfers, for a determined period of time, to a Romanian or foreign legal person, as grantee, on the one hand, the right and obligation to perform, on its own risk and expense, the petroleum operations that fall under the Petroleum Law, in return for a royalty and, on the other hand, the right to use public property goods, which are necessary for the performance of the petroleum operations, in return for a royalty'.
380. According to Art. 2, point 2 of Petroleum Law, a petroleum agreement is 'the legal document concluded under the Petroleum Law between the competent authority and one or more Romanian or foreign legal entities, for granting an oil concession'.

authority, NAMR, as 'lessor' or 'grantor'[381] and a Romanian or foreign legal person, as 'lessee', 'grantee' or 'title holder'.[382] All petroleum agreements must be approved by Government Decision.[383]

381. *See* Art. 2, point 6 corroborated with Art. 2, point 3 of Petroleum Law.
382. *See* Art. 2, point 8 corroborated with Art. 40 of Petroleum Law.
383. *See* Art. 22 para. 1, letter a) of the Rules.

Chapter 3. Concession Agreements

325. Under a petroleum agreement the state grants to the title holder either a non-exclusive[384] exploration[385] right (exploration license), or an exclusive development and production right (lease or concession) for each commercial discovery,[386] within a determined perimeter.[387]

§1. General Regime of Concession Agreements

326. Under the Constitution,[388] all subsoil mineral resources of public interest are exclusive public property and their exploitation can be granted or leased in accordance with the organic law. Otherwise, the property rights over mineral resources are inalienable.

327. The general regime applicable to the content, exercise, and defence of concession rights is governed by the NCC. With respect to the *content,* the NCC[389] states the grantee has both the right and the obligation to exploit the granted goods, in exchange for a royalty, for a determined period of time and with the full observance of legal conditions and contractual provisions. Both Romanian and foreign legal or natural persons[390] can be grantees.[391] However, the concession procedure as well as the conditions required for the conclusion, performance and termination of a concession agreement are to be governed by law. This means that the Civil Code's framework is insufficient for the conclusion of concession agreements,[392] in the absence of a special law to govern the three aspects mentioned.

384. *See* Art. 2, point 27 of Petroleum Law.
385. *See* Art. 2, point 15 of Petroleum Law.
386. *See* Art. 5(2) of Petroleum Law corroborated with Arts 71(1) and 72 of the Rules.
387. *See* Art. 3 of Petroleum Law. Also *see*: Arts 21 and 57 of the Rules.
388. Article 136(3)–(5).
389. Article 871 of the NCC.
390. In certain cases natural persons cannot acquire concession rights, as it is the case for petroleum operations. *See*: Art. 2, point 8 of Petroleum Law and Valeriu Stoica – Drept civil. Drepturile reale principale. Vol 1 (Civil Law. Main Real Rights. Vol. 1), Ed. Humanitas, Bucuresti, 2004, p. 450.
391. The current provisions do not distinguish between private or public legal persons as the previous legislation did. Hence, under current Romanian legislation, public legal persons can acquire concession rights.
392. Colectiv – Noul Cod civil – comentarii, doctrina, jurisprudenta – Vol I (The NCC – Comments, Doctrine and Case Law – Vol I), Ed. Hamangiu, Bucuresti, 2012, p. 1207.

328. Despite the fact that the general rules are contained in the NCC, the concession right is deemed by the doctrine[393] and by courts[394] to be an *administrative one*, given the fact that the grantor may, for reasons related to public interest and with adequate compensation, unilaterally terminate the contract and recover the granted goods. As such, the concession right holds the following legal characteristics:[395] it contains a *prerogative of exploitation* both in public and personal interest, exploitation is *mandatory*, under the risk of losing the concession and it is always *onerous*, for the payment of royalty is of the essence of the concession agreement.

329. With respect to the *exercise* of the concession rights, the NCC[396] establishes that the grantee is entitled to perform any legal or material acts which are necessary for the exploitation of the goods. However, under the threat of absolute nullity, the grantee may not dispose of or encumber the conceded goods or, respectively, those goods destined or resulted from the exploitation of the concession, and which, according to the law or the granting document[397] must be returned to the grantor at the termination of the concession.[398] The sanction of nullity derives obviously from the inalienability rule established in the Constitution. The doctrine[399] emphasized that not only the disposition or encumbrance of granted goods is covered by this sanction, but also those concerning mandatory investments or own goods, which, according to the concession agreement, are to become the property of the grantor at the end of the concession under a takeover clause. However, in certain circumstances the transfer or assignment of acquired rights may take place, but only with the prior approval of the grantor, as in the case of farm-out agreements.

330. Fruits and, within the limits of the law or the agreement, products are the property of the grantee. If in case of fruits the issue is rather simple, because the

---

393. Id at 1206 and Stoica, p. 450. The latter emphasized that the specificity of the concession agreement consists of not only the co-existence of a so-called *normative part*, containing mandatory provisions, notwithstanding whether they were specifically included in the contract or not, with a per se *contractual part*, but also in the fact that the general rules applicable to civil contracts do not govern the entirety of the concession agreement, but only the contractual part, for it is the only part which has been negotiated by the parties and represents their common will.
394. *See* Decision No. 6/2011 of the High Court of Cassation and Justice holding that the concession agreement concerning the exploitation of public property goods is assimilated to administrative acts as defined by Art. 2(1) c) of Administrative Litigation Law No. 554/2004. Available online, in Romanian, at: http://www.scj.ro/SCA%20rezumate%202011/SCA%20dec%20r%206%202011.htm.
395. Colectiv, at 1206–1207. According to Stoica, p. 450, the characteristics of a concession contract are: *intuitu personae,* solemn, mutually binding (synallagmatic), onerous, commutative, with successive performance and constitutive of real rights (in rem rights), while the characteristics of the concession rights are: inalienability, unencumbered, indefeasible, but temporary (p. 452).
396. Article 872 of the NCC.
397. The granting document is normally the concession agreement. However, the general law does not exclude here the possibility that the granting document is an administrative unilateral act, such as a license for prospections (in case of petroleum operations). *See* Art. 2, point 27 of the Petroleum Law.
398. Stoica, p. 448.
399. Colectiv – Noul Cod civil – comentarii, doctrina, jurisprudenta – Vol I (The NCC – Comments, Doctrine and Case Law – Vol I), Ed. Hamangiu, Bucuresti, 2012, p. 1207–1208. Also Stoica, p. 449.

right to exploit the granted goods automatically includes their fruits (be it natural, industrial or civil[400]); in case of products, which affect the very substance of the granted goods, there must be an expressed clause attesting that they belong to the grantee. Such clauses are of essence in petroleum or mining concessions.

331. The exercise of concession rights is not an absolute prerogative, since the grantor is entitled to control it, under the provisions of the law and the concession agreement. The right to control represents, together with the right to unilateral termination, one of the specificities of administrative contracts, in which parties do not enjoy contractual equality.[401]

332. Concerning the judicial *defence* of the concession rights, the NCC[402] provides that this obligation belongs to the grantee. The doctrine[403] takes the view that given the mandatory aspect of the exploitation of the granted goods, it makes sense that the defence of concession rights should be exercised by the grantee, under the sanction of rescission for breach of contractual obligations.[404] In case the concession rights cannot be exercised due to the grantor's actions, the grantee is entitled to bring judicial action against the grantor as well.

333. The jurisdiction for such defence actions belongs to the general court, notwithstanding who the aggrieving party is, unless the disturbance was done by a public authority or through administrative acts, in which case the jurisdiction belongs to the Administrative Litigation Courts.[405]

334. The grantee cannot exercise such defence against the grantor, where the latter has decided to unilaterally terminate the contract for public interest reasons, unless it was exercised with excess of power, for which the grantee is entitled to bring a separate action under the Administrative Litigation Law.[406]

335. The concession agreement may be terminated:[407] by the expiration of its term, unilaterally by the grantor, in case national or local interest demands it and with adequate compensation, for breach of contractual obligations, by either of the parties, with payment of damages and by the disappearance, for force majeure reasons, of the granted goods or in case its exploitation becomes objectively impossible, without payment of damages. In each case, the grantee must return the goods free of any encumbrances.

400. In case of civil goods deriving from rents or interests one must ensure that they do not result from subleases for they are banned by the law. *See* Art. 51 para. 3 of GEO No. 54/2006.
401. Colectiv, p. 1208. Stoica, p. 456.
402. Article 873 of the NCC.
403. Colectiv – Noul Cod civil – comentarii, doctrina, jurisprudenta – Vol I (The NCC – Comments, Doctrine and Case Law – Vol I), Ed. Hamangiu, Bucuresti, 2012, p. 1208.
404. *See* Art. 57 para. 1, letter c) of GEO No. 54/2006 concerning the concession contracts' regime.
405. *See* Art. 66 of GEO No. 54/2006 concerning the concession contracts' regime.
406. *See* Art. 1 of Administrative Litigation Law No. 554/2004.
407. Article 57 of G.E.O. No. 54/2006 concerning the concession contracts' regime.

336. The above-mentioned termination causes are mandatory and non-waivable for the parties, but not limitative,[408] which means the parties are free to negotiate and add other causes as well. With respect to the right of unilateral termination, one may notice that although within the powers of the grantor, it cannot be exercised arbitrarily, but only if legitimate public interest demands it. The appreciation of such legitimacy is left up to the courts. With respect to termination for breach, the doctrine took the view that, although according to Romanian law, as a general rule it has a judiciary character, nothing precludes the parties to include an express right for extra-judicial termination (pact comisoriu).[409] Concerning the 'objectively impossible' formula, the doctrine stated that along with force majeure causes it also includes unforeseen circumstances.[410]

§2. Special Regime of Concession Agreements under the Petroleum Law

337. Concession agreements concerning petroleum operations together with the public property goods necessary for their performance are governed by the special regime established by the Petroleum Law together with the Rules. One of the first special rules regards the duration of the concession which is limited to thirty years, with the possibility of prolongation for additional fifteen.[411] Therefore, the maximum term for a petroleum concession according to Romanian law is forty-five years.

338. The list of available perimeters for petroleum operations is established by Order of NAMR which is published in the Official Gazette of Romania and in the Official Journal of the European Union.[412]

339. The petroleum agreements must be concluded in writing and come into force after their approval by the Government. Modification is allowed by addendums, which must also be approved by the Government. Additionally, the beginning of the performance of petroleum operations must be authorized, in writing, by the competent authority (NAMR), on the basis of authorizations and approvals required by applicable law.[413]

340. Foreign legal persons that were granted the right to conduct petroleum operations are obliged in ninety days from the coming into force of the petroleum agreement, to establish and maintain a branch or a subsidiary in Romania for the entire duration of the petroleum operations. In case the concession was granted to

408. Stoica, p. 454.
409. *Ibid.* at 455.
410. *Ibid.* at 455.
411. Article 27(2) of Petroleum Law.
412. Article 29 of Petroleum Law.
413. Article 31, paras 1–3 corroborated with Art. 32 of Petroleum Law.

more than one foreign legal entity, the aforementioned obligation lies with the company designated to represent their interests.[414]

I. Prospection Works

341. However, not all petroleum operations require the conclusion of a concession agreement. *Prospection works* can be performed also on the basis of *a non-exclusive license* for prospection, granted by NAMR on written request, for a defined perimeter.[415] The license must be issued in thirty days from the interested party's written request (provided all other legal conditions are fulfilled[416]), it is valid for three years only and cannot be prolonged.[417] Unlike petroleum agreements, licenses do not need to be approved by Government Decision.[418] In case the holder of the license wishes to get a concession for exploration of the prospected perimeter, then he must submit, within sixty days, together with the concession request, a report containing the investigation methods utilized, the justification of his expenses, as well as the results obtained.

II. Exploration Works

342. Exploration works[419] designate the set of studies and operations conducted for knowing of geological conditions for oil accumulation and it includes the identification of deposits, their quantitative and qualitative evaluation, as well as the determination of the technical and economic conditions for their recovery. Their purpose[420] is to substantiate any decision concerning the opportunity of exploiting the deposit as well as to provide the necessary data concerning the projection and performance of opening, preparation and exploitation works, while maintaining a degree of insurance with adequate oil resources and reserves for exploitation.

343. Provided that petroleum deposits are discovered, the deposit may be declared commercial deposit. Following the first well with results, the holder of the concession or license will elaborate an evaluation project of the deposit by detailed geophysical works and/or through additional wells. The evaluation project of the discovery must be approved by the NAMR.[421]

414. Article 33 of Petroleum Law.
415. The shape and dimensions of the perimeter are determined by the competent authority. *See*: Art. 21(2) of Petroleum Law and Art. 28(2) of the Rules.
416. The required conditions are enumerated in Art. 57 of the Rules.
417. Article 28(1) and (3) of Petroleum Law.
418. *See* the provisions of Art. 22(3) a) and of Art. 22(2) of the Rules.
419. Article 2, point 15 corroborated with Art. 60(1) of the Rules. According to the provisions of Art. 60(3) they include: geochemical and geophysical works, geological mapping, exploration drilling – opening and exploration – evaluation, their corresponding documentation, experimental exploration and scientific research. Scientific research and identification of deposits are further defined in Art. 61, respectively 62 of the Rules.
420. Article 60(2) of the Rules.
421. Article 63(1) and (2) of the Rules.

344. According to the Rules,[422] the petroleum agreements will establish the volume and the value of the minimal works programme, necessary for the determination of the petroleum perspectives in the perimeter, together with the schedule for performance. Any additional works and schedules are established by direct negotiation with the NAMR and are approved through the petroleum agreement.

345. Both minimal and additional works are performed on the basis of projects elaborated by the holder with the approval of the agency. The same conditions[423] will apply to the experimental exploitation, under the conditions stipulated in the agreement. However, if following the experimental exploitation, it is determined that the deposit cannot be commercially developed further, the holder is entitled to solicit an exploitation permit through the exploration well, for as long as he deems this process feasible from both an economic and technologic standpoint.

346. The part of the granted exploration perimeter which, following performed works, did not reveal any commercial deposits according to the concession agreement, is to be relinquished to NAMR free of any encumbrances. Normally, both the renunciation term and the relinquishment date at which the exploration period ends take place according to the concession agreement, unless intermediary terms were provided for.[424]

III. The Concession Agreement: Initiation

347. According to Romanian law, the concession of petroleum operations may be initiated either by national or foreign legal persons, or by the NAMR. In the case of the former, those interested may initiate the concession by sending a written request to NAMR, together with the chart of the perimeter, elaborated according to the technical norms,[425] while in case of the latter the interested parties must answer to a public call for tenders.

348. Following the written request, NAMR verifies and establishes the perimeter's area, its coordinates and decides upon the concession's suitability within forty-five days from the receipt of the request. The decision is then communicated to the applicant within five days from its issuance.[426]

349. The list of perimeters which is publicly offered for concession are approved by Order of the head of NAMR[427] and are published in the Official Gazette of

422. Articles 65 and 66 of the Rules.
423. Article 68 of the Rules.
424. Article 70 of the Rules.
425. Articles 27–28 of the Rules.
426. Article 29 of the Rules.
427. According to Art. 31 of the Rules, the Order contains the following elements: the name and headquarters of NAMR, the category of petroleum operations to be concessioned, the perimeters and

Romania and the Official Gazette of the European Union, together with the criteria for evaluation of the bids and of the technical and financial capabilities of the bidders.[428]

350. Interested parties shall submit their bids in a term which cannot be shorter than 30 days or longer than 270 days, as established by Order of NAMR.[429] According to the Rules[430] the offers shall be submitted in sealed envelopes[431] and shall be registered in the order of their receipt. Until the public session of bids' opening, they shall be kept sealed in full confidentiality. Any bid received after the deadlines shall be returned unopened and unregistered to the bidder.

351. According to the Law,[432] the bids[433] are written in Romanian[434] and contain: (1) the proposed exploration programme;[435] (2) the proposed development-exploitation or solely exploitation programme,[436] as the case may be, (3) any other documents requested by the NAMR.

352. The conditions for organizing and conducting the public call for tender as well as the selection and winning bidder designation criteria (together with any

their coordinates, the term within which the interested parties may obtain the procedure of the public call for tender, the deadline set for concluding the concession agreement and any other information deemed necessary.

428. Article 30 of the Rules.
429. Article 29, (4) of Petroleum Law.
430. Articles 36–38 of the Rules.
431. According to Arts 34 and 35 of the Rules, the offer shall contain to envelopes: one external and one internal, containing the requested data and documents contained in the Rules and the public call for tenders. The *exterior envelope* shall indicate the perimeter for which the bid is submitted and the established date for the opening session. This exterior envelope must contain: a clear and express declaration of participation, specific information concerning the bidder, certified copies of the bidder's articles of incorporation, tax number, social capital, shareholders and its legal representative, a certificate issued by the Trade Register concerning national bidders or any similar document concerning foreign bidder, a list of data and information used to elaborate the bid, a creditworthiness letter issued by the bidder's bank, concerning its solvability, and certificates issued by the public institutions administering the public budgets, concerning the bidder's financial file, or in case of foreign bidders, the final annual audit report of the bidder. The *internal envelope* shall indicate the name and the headquarter of the bidder and shall contain the actual offer, respectively the proposed petroleum operations, an estimation of reserves to be discovered or pursued, a cost evaluation of the operations, the operations stages, an estimate of the environmental impact and the restoration programme, the continuous training programme of the employees and the technology transfer programme.
432. Article 29(5) of Petroleum Law.
433. According to Art. 33 of the Rules, one cannot submit more than a bid for each perimeter.
434. Article 32 of the Rules. The same article establishes an obligation for the bidders to submit duplicates: an original and a copy.
435. According to Art. 29(6) of the Petroleum Law, the programme shall contain a description of mandatory petroleum operations to be performed and their estimated costs.
436. According to Art. 29(7) of the Petroleum Law they shall contain the estimated reserves, the period of economic capitalization of the reserves and the development works proposed.

other aspects) are established through transparent and non-discriminatory government approved rules, which shall derogate from the general legal framework concerning the concession's regime.[437]

IV. The Concession Agreement: The Opening, Evaluation and Negotiation Committees

353. For each public call for tenders, the NAMR's head appoints by Order the committees for opening and evaluating the bids as well as for negotiating the conditions and the provisions of the petroleum agreements.

354. The *Committees* consist of five members and two alternate members, all representing NAMR. The committees shall be deemed legally formed only in the presence of all members[438] and shall decide with a majority of votes.[439] Their management is ensured by a president appointed by NAMR's head. The members of the Opening Committee may serve as members in the Evaluating Committee as well.[440] However, those who are spouses, kin or close up to the 4th degree inclusively, to the shareholders or partners, administrators or censors of the bidder, are precluded from being members[441] in any of the committees. Related objections can be raised to the president of the committee by any interested person.[442]

355. The Opening Committee has the following main duties:[443] (a) to validate the submitted bids; (b) to open and designate the bids admitted for evaluation; (c) to prepare the minutes of the opening meeting; (d) to submit the minute to the NAMR's head for validation; (e) to communicate to the bidder a copy of the minute; and (f) to transmit to the Evaluation and Negotiation Committee the admitted bids.

356. The Evaluation and Negotiation Committee has the following main duties:[444] (a) to evaluate the bid, according to the criteria established by the rules of organization of the public call for tenders; (b) to designate the winning bid; (c) to prepare the minute concerning the decision made; (d) to submit the minute to the NAMR's head for validation; (e) to notify all the bidders of the evaluation's result; (f) to communicate to the bidder a copy of the minute; (g) to negotiate the provisions and the conditions of the concession; (h) to draft up the petroleum agreement, initialize it and submit it for signatures to both the NAMR's head and to the legal representative of the winner, and (i) any other duties established by NAMR's head.

---

437. Article 29(8) of Petroleum Law.
438. Article 46(1) of the Rules. In case one of the members cannot participate, he/she is replaced by one of the alternate members.
439. Article 46(3) of the Rules. Dissenting opinions are recorded in the meeting's minute.
440. Articles 40 and 41 of the Rules.
441. In this regard, according to Art. 43(2) of the Rules, each member of the committees must submit a declaration concerning its compatibility, on his/her own recognizance.
442. Article 43 of the Rules.
443. Article 44 of the Rules.
444. Article 45 of the Rules.

V. The Concession Agreement: The Selection Criteria and the Corresponding Scores

357. As already mentioned, the Evaluation and Negotiation Committee evaluates the bids based on the criteria specified in the rules and clarifications concerning the procedure for each public call for tender.[445] The Rules establish detailed scores for each petroleum operation – exploration, development and exploitation, respectively for development, exploitation or solely exploitation. The criteria employed refer to the minimum programme proposed, the bidders' technical and financial capacities, the impact on the environment and the programme for restoring it, and the technology transfer and continuous training of personnel programmes.[446]

358. The scores are so set that the total number of points is 100. The evaluation is done by scoring every single evaluation criteria. Any offer scoring less than half of the maximum number of points is automatically disqualified. In case of tie scores, the winning bid is the one with the highest score for the criteria concerning first stage minimum mandatory works.[447]

VI. The Concession Agreement: The Opening of the Bids

359. The bids are opened in a public session. All sealed bids are handed over to the Committee at the date, time and place established in the procedure for the public call for tenders. Upon opening the external envelope, the committee verifies the presence of the requested documents and records the situation in a minute signed by the members of the committee and the bidder's representatives.[448] The absence of any requested document results in automatic disqualification of the offer, without the opening of the internal envelope. In case all documents are present, then the internal envelope is opened, with the situation recorded in a minute, similar to the procedure concerning the external one. The admitted bids are immediately transmitted to the Evaluation and Negotiation Committee. The evaluation of the bid is also recorded in a minute which is communicated to the bidder.[449]

VII. The Concession Agreement: Challenging the Public Call for Tenders

360. Bidders may file disputes about the way in which legal provisions concerning the organization and the performance of the public call for tenders took place, both with respect to the opening or the evaluation of bids, within five working days from the receipt of the minute validating the decision of the commissions. Within

445. Article 48 and 45 (a) of the Rules.
446. Article 49 of the Rules.
447. Article 50 of the Rules.
448. In case the bidder's representatives are not present or they refuse to sign the minute, then the minute is signed by the committee's members and two witnesses from the persons attending the public opening session, other than NAMR's representatives.
449. Article 51 of the Rules.

ten days from receiving the dispute, the NAMR must solve it and communicate its answer to the challenging bidder.[450] The dispute suspends the concession procedure until a final and definitive solution is reached.[451]

361. Disputes are solved by a Dispute Committee, formed of three members of NAMR who have not been part of the Opening or the Evaluation Committees. All three members are appointed by decision of the NAMR's head.[452]

362. The Dispute Committee analyses all documents issued by the Opening or, respectively, the Evaluation Committee and verifies whether the legal procedure concerning the public call for tenders was followed. Upon the analysis of the contested facts and based on the applicable legal provisions, the Dispute Committee prepares a report which is submitted for approval to the NAMR's head, by which it proposes a solution to solve the dispute.[453]

363. Where the challenge is well founded, the committee proposes to the NAMR's head the following, as the case may be: to revoke the decision of disqualification (where the dispute concerns the opening phase) or to revoke the decision designating the winner and to organize a new public call for tenders (where the dispute concerns the evaluation phase). The Dispute Committee may not re-evaluate the bids.[454]

364. The decision to uphold the challenge is communicated to all bidders that have placed tenders for the perimeter in question. *Per a contrario,* the decision to dismiss the challenge is communicated only to the bidder that raised it, who is then free to contest it in front of the court having jurisdiction. In case of court proceedings, the concession is also suspended until a final decision is rendered, however, the claimant must pay a surety in the amount set by the court.[455]

VIII. Termination of Concession Agreements and Its Effects

365. A concession is terminated when: (a) it reached the term for which it was granted; (b) if the holder of the petroleum agreement relinquishes it under the conditions established by law; (c) by NAMR's unilateral termination under the conditions established by law[456] and (d) at the request of the holder, where force majeure

450. Articles 52–53 of the Rules.
451. Article 55, para. 6 of the Rules.
452. Article 54 of the Rules.
453. Article 55, para. 1 corroborated with Art. 56 of the Rules.
454. Article 55, paras 2–3 of the Rules.
455. Article 55, paras 4–7 of the Rules.
456. The full list of reasons for unilateral termination by NAMR is contained in Art. 42 of Petroleum Law and includes: non-fulfilment by the license holder of the undertaken obligations concerning the start of the petroleum obligations at the agreed term; the license holder suspends the petroleum operations for more than sixty days without the prior approval of the authority; the holder performs

events preclude the fulfilment of the holder's rights and obligations as set out in the concession agreement and which are deemed as essential for the performance of the petroleum operations.[457]

366. The decision concerning the termination of the concession agreement is published in the Official Gazette of Romania and must mention the date from which it comes into effect. The decision is also mentioned in the Petroleum Book.[458]

367. At the date of termination, the license holder is obliged to return, in full ownership, all the leased goods, free of any encumbrances, including all the investments resulting from the performance of the petroleum operations.[459]

---

unauthorized petroleum operations, the holder willingly provides false data concerning the petroleum operations to NAMR, the holder does not pay the royalty for six months, etc.

457. Article 38 of the Petroleum Law.

458. Article 44, para. 1 of Petroleum Law.

459. Article 44 para. 2 of Petroleum Law.

Chapter 4. The Farm-Out Agreement (Transfer of Rights and Obligations Derived from Petroleum Agreements)

368. Petroleum operations can be carried out directly or indirectly (through subcontractors) by the holder or they may opt to transfer, fully or partially, the rights and obligations of the Agreement to a third party. The contract by which such wholly or partly transfer of rights, resulting from a petroleum agreement, is made, is generically named FARM-OUT.

369. Farm-out Agreement is an unnamed contract in Romanian legislation, its title originating in Anglo-Saxon law. In Romania, it is governed by Article 34 of Petroleum Law corroborated with the provisions of Articles 97–101 of the Rules.

370. Given that it is an unnamed contract, there is no legal definition for this type of contract,[460] being necessary an attempt to have it defined by the doctrine, based on legal provisions and by reference to international doctrine. Thus, strictly based on legal texts, one can define the farm-out agreement as a contract whereby the holder of a petroleum agreement (the Farmor) transfers, in whole or in part, to another legal person (called Farmee) its rights and obligations, only with the prior written approval of NAMR and with the cumulative fulfilment of all the conditions stipulated by law.

371. From the above definition one can thus distinguish the legal characters of the farm-out agreement: it is a bilateral contract, solemn, *sinalagmatic* (mutually binding), onerous (for a consideration), mixed.

372. It is a *bilateral* contract because, as stated in the law, it may be concluded between the holder of the petroleum agreement and another legal person, Romanian or foreign, who is both willing and able (by law) to undertake the rights and obligations arising from the petroleum agreement. Nothing prevents the holder (Farmor) to transfer its rights and obligations under the concession agreement to more interested legal persons (Farmees). This, however, may only happen through contracts signed separately with each Farmee.[461]

373. It is a *solemn* contract because its validity depends on the prior written permission of NAMR, and in its absence, the contract is null and void. NAMR's written consent is manifested by an Order issued by the head of the Agency.[462]

374. It is a *sinalagmatic* (*mutually binding*) contract because, once signed, it is binding for both parties, their obligations are mutual and interdependent, each one of them having their cause in the obligation of the other.

---

460. The closest form of a definition is the explanation contained in Art. 97 para. 1 of the Rules.
461. However, the joint operation agreement (JOA) will include all holders of rights and obligations arising from the oil concession agreement.
462. *See* the provisions of Art. 101 of the Rules.

375. It is an *onerous* contract because each party pursues its own economic purposes. The Farmor recovers all or part of the investment made in obtaining the concession and possibly, depending on the stage of operations at the time of concluding the farm-out contract, the costs incurred by these operations, but also gets the financial participation of the Farmee in the operations to be performed after completion of the transfer. Meanwhile, the Farmee obtains its participation in an existing petroleum agreement, shares operations costs that will be incurred from the time of the transfer and obtains, depending on the contractual provisions, equity interests in any discoveries that will be made.

376. It is a *mixed* contract because the transfer involves, on the one hand, only the definite existence of a petroleum agreement and the amount to be paid in advance by Farmee, and, on the other hand, the minimum amounts that will be invested, but there is no certainty concerning the discoveries that will be made, the total amounts to be invested or the profits realized, for these will result later, from further events which lie beyond the effective control of the parties. The farm-out agreement, presents in this respect, both features of commutative contracts and random contracts, for which reasons this monograph deems it a mixed contract.

§1. Conditions of Validity

377. To achieve the transfer of rights and obligations arising from the petroleum agreement, Romanian legislation imposed a number of conditions that must be fulfilled in advance of concluding the farm-out contract. There are also conditions arising from legal forms, without being expressly referred to by law as validity conditions. The general conditions established by the NCC must also be observed, but these general conditions are not part of the following analysis.

I. Conditions Concerning the Farmee

378. The law states that one may transfer the rights and obligations contained in the oil agreement to any other legal person(s).[463] The text does not mention if the legal person(s) may be national or foreign, as it did in the case of holders of agreements, but since this is a transfer of rights and obligations emanating from such an agreement, it may be safe to assume that the legislature has not considered necessary to state the same provision twice. It is obvious that as long as the holder of a concession agreement may be a Romanian or foreign legal entity, the one to whom the transfer is being made can also be a Romanian or foreign legal entity. Following the same logic by which the holder (the Farmor) and the entity to which the transfer is being made to (the Farmee) transmit rights and obligations which they have undertaken from and towards the Roman state, it follows that in the event that the Farmee is a foreign legal entity, it will have to establish and maintain a subsidiary and a branch in Romania.

---

463. *See* both Art. 31 para. 1 of Law and Art. 97 of the Rules.

379. In order to seek NAMR's approval for the transfer, the Farmee must first prove its technical and financial capacity needed to perform the petroleum operations in the conditions determined by agreement.[464] The Farmee must also be a legal entity specialized in carrying petroleum operations or, at least designate as Operator an authorized company who possess adequate technical capacity for the petroleum operations arising under the agreement.[465] This is because one of the common obligations of the Farmor and the Farmee is to maintain the existing conditions of the license, as set out in the concession agreement[466] unless, after negotiations with the competent authority, other obligations have been established by addendum.[467]

II. Conditions Concerning the Farmor

380. The Farmee must meet one main obligation: to be the holder of the agreement,[468] and the agreement must still be in force,[469] implying that the holder (the Farmor) has fulfilled all obligations arising under the petroleum agreement until the transfer takes place.[470] From this last condition the law established one exception, namely that in the event that not all obligations have been met, then the legal entity to which the transfer is being made (the Farmee) expressly agrees to take over the unfulfilled obligations. Fulfilment or non-fulfilment of obligations undertaken must be apparent from a report on the operations of the holder (Farmor) until the transfer request. Preparing and submitting the report to NAMR falls within the duties of the Farmor.

381. At the time of the transfer, the Farmor must also meet and maintain the common condition to maintain the existing conditions of the concession, as was established by the petroleum agreement, if after negotiations with the competent authority other conditions have not been established by addendum.

III. Conditions and Authorization Procedure

382. Before the formal transfer of rights and obligations of the Farmor to the Farmee, it is necessary to obtain a written agreement to this effect from NAMR.[471]

464. *See* Art. 34 para. 3 of the Law and Art. 98 letter d) of the Rules.
465. *See* Art. 98 letter f) of the Rules.
466. *See* Art. 98 letter e) of the Rules.
467. In accordance with Art. 100 para. 2 of the Rules, by agreement of the parties some terms of the petroleum agreement may change by signing an addendum between the new holder and NAMR.
468. *See* Art. 34 para. 1 of the Law and Art. 97 para. 1 of the Rules.
469. *See* Art. 98 letter a) of the Rules.
470. *See* Art. 98 letter c) 1st thesis of the Rules.
471. *See* the corroborated provisions of Art. 34 of the Act ('the holder can transfer .. only with prior written approval') with those of Art. 97 para. 1 of the Rules.

NAMR's approval, which cannot be refused without good reason, is a powerful formal restriction of the right of a holder of a petroleum concession agreement to exercise his transfer option, and seeks to protect in particular the rights of the Romanian state, in its grantor capacity.

383. Thus, by checking Farmee's possibility to fulfil the conditions of the petroleum agreement the state seeks to ensure that the conditions stated in the public tender offer that lead to the determination of a winner according to the law are not circumvented The grantor also ensures that it knows who is/are the holder(s) of the petroleum agreement at any time. Meanwhile, the rights of third parties are protected, for NAMR's approval fulfils an advertising role and, last but not least, the Farmee's rights are protected for thus it can find out whether the Farmor has fulfilled its obligations.

384. The legal consequence of the lack of a NAMR's prior approval is automatic nullity of the transfer.[472] Under the terms of NAMR's standard terms of oil agreements, this approval cannot be arbitrarily denied by the agency, but only when the conditions stipulated by the law are not met.

385. In order to obtain the said approval, the Farmor and the Farmee must address, in writing, the competent authority with a joint application, signed and stamped by the legal representatives of the two legal entities. The application must be accompanied by a series of documents,[473] as follows:

- A certificate issued by the Trade Register concerning the legal person to which the petroleum agreement rights and obligations are to be transferred or an equivalent of such certificate, in case of foreign firms.
- The decisions of the governing bodies of the two legal entities, which approve the transfer of rights acquired and the obligations assumed.
- A certified copy of the Articles of Incorporation and the certificate of registration of the legal entity to which the rights and obligations are to be transferred.
- A report on operations carried out by the holder until the request for approval of the transfer of rights acquired and obligations assumed, with express mention of the status of fulfilment of obligations arising from the concession agreement and the results obtained.

386. The agency has thirty days to consider and decide on the request and, in the event that the legal conditions are fulfilled, all requested documents were submitted, and where not all obligations have been met but the legal entity which the transfer is being made (Farmee) expressly agreed to take them over, the head of NAMR approves the transfer by order.

472. Article 34 para. 2 of the Law provides only that 'any transfer made without the approval of the competent authority is null'. The indication that the nullity is automatic (*by law*), derives from the wording of Art. 97 para. 2 of the Rules 'the transfer made without NAMR is *by law* null and void'.
473. *See* Art. 99 para. 2 of the Rules.

§2. THE OBJECT OF THE TRANSFER

387. According to Article 34 of the Petroleum Law, the license holder can transfer totally or partially the rights and obligations acquired. The law now clarifies that in the case of a partial transfer, this can refer to: a share of all the rights and obligations acquired through the concession agreement for the entire perimeter; a share of all the rights and obligations acquired through the concession agreement for a certain area; or the entirety of all the right and obligations acquired through the concession agreement for a certain area.[474] In the last two cases, the competent authority may set out the boundaries of the areas within a petroleum perimeter, under the procedure established by Article 34^1 of Petroleum Law.

388. Specifying the exact equity interest transferred is essential for determining the proper form necessary to make the transfer.

389. As already shown, in Romania, minerals and natural resources of the subsoil are exclusive public properties, are inalienable and can only be leased under the law. The regime of concession of public property is currently regulated by the NCC which states *inter alia* that the concession holder – given his status – may perform any material or legal acts necessary to ensure good operation of the leased goods.

390. It is possible that by the time of transfer, the holder of the petroleum agreement had already performed surface and subsoil works, which in their turn can constitute or not, an object of the transfer. In this category of tangibles enter for example: tubes, pipes, pumps, compressors, tanks, oil tanks, the equipment used to operate the wells but also buildings.

391. The transfer of goods mentioned above can be done either through a provision in the farm-out agreement or by a separate deed, but regardless of the method chosen, it is obvious that in case the transferred goods include buildings then the deed must comply with the formal general requirements of civil law concerning transfers of immovable goods. In my opinion the transfer of such goods by farm-out agreement is valid only if they are expressly mentioned in the contract, otherwise it will be presumed that they were not subject to agreement of the parties.

392. It must be noted here that not only the rights and obligations contained in the petroleum agreement may be subject to farm-out agreement, but also other rights and obligations agreed by the parties.

393. Also, it is necessary to define precisely the rights transmitted. Although the law does not distinguish, this monograph takes the view that not only a total or partial share of the entire object of the petroleum agreement may be transferred, but the parties are free to define the object according to their own interests.

---

474. Article 34, para. 1^1-1^2 of Petroleum Law.

394. Thus, it is possible to transfer only certain areas of land from the entire leased area, provided it is precisely identified and determined in order to avoid possible location problems. One can also transfer a specific well (or more), the subsoil from or up to a certain depth, in the latter case being necessary an expertise carried out by professionals in the field to precisely define what is transferred.[475]

395. A problem related to the definition of the object of the contract derives from multiple transfers. Suppose A transfers to B all its rights, retaining for himself a share of 1/5 of the total oil and gas production. In turn, B transfers to C ½ of all its rights. As C is not part of the agreement between A and B it is possible that C might consider it acquired ½ of the entire future production of oil and gas, when in reality he only gained half of what has remained to B, meaning 4/5 of the total initial rights. This monograph is of the opinion that under the provisions of Romanian law, questions of interpretation are easily resolved, on the one hand, because B cannot transfer more rights than it has, and, on the other, because at the time of the transfer from B to C, a validity check of B's title – which is the initial transfer – will occur. Even so, provided the contract does not state otherwise, one might argue that B assumed to cover for the rights which belong to A, by failing to state precisely the object in the agreement to transfer the rights to C. Therefore, defining exactly what is transmitted through careful and precise drafting of the farm-out agreement can avoid such problems.

§3. The Rights and Obligations of the Parties

396. This section addresses only the rights and obligations arising from the farm-out contract after having received approval from NAMR, for the obligations of the Farmor and the Farmee normally include those required to be met before approval, which were addressed in the previous section. But unlike the aforementioned obligations, which arise from the law, the obligations mentioned in this section are purely contractual. As such, they vary from one case to another, depending on what the parties negotiated and agreed on. It must be mentioned that although simple in appearance, the rights and obligations arising from the farm-out contract are in fact complex.

I. Rights and Obligations of the Farmor

397. The Farmor is entitled to receive the amount established (if such amount was set) in the contract in exchange for the equity interest it transferred (be it total or partial). It is also entitled to all the benefits resulting from its equity interests kept, in line with the contractual provisions.

---

475. The problems of measuring the depth depend on the method of measurement chosen: vertical depth or well depth. Given that deviations often occur during drilling, it is possible that the well depth is greater than the vertical depth so, in order to avoid any misunderstandings, an expert's advice is recommended before drafting the contract.

398. Regarding its obligations, the Farmor will have to pay the contribution that accrues from all works to be carried out. Typically the share that accrues is equal to its remaining equity interests although other types of agreements can be negotiated.[476]

II. Rights and Obligations of the Farmee

399. Once it made payment, the Farmee is entitled to enjoy all rights under the equity interests acquired. An interesting question concerns the moment when the Farmee becomes the holder of the equity interests, in cases where, in addition to an initial lump sum, the Farmee assumed also the payment or performance of future works. From the wording of the studied contracts, it results that the transfer occurs immediately, after signing the contract and obtaining NAMR's approval. In this situation, one might conclude that Romanian law's position is that the payment condition is purely contractual, one which is not of the essence of the farm-out agreement and would affect the mutually binding aspect of the contract. However, my opinion is that in reality, the transfer of share interests resulting from a petroleum agreement occurs under a resolutory condition. Hence, it follows that the Farmee will enjoy the fullness of its rights only in the event that it meets all its obligations, at the agreed terms, during the performance of the contract.

400. At the same time, it should be mentioned that together with the rights expressly mentioned in the Farm-out agreement, the Farmee will benefit implicitly from the rights resulting from it and from the oil concession agreement, such as for example the right to perform surface works, to use public utilities, pipelines, ports and other facilities needed for oil operations, to use surface or underground water sources, build roads and railways needed, to seek the extension of the leased perimeter or to associate with other legal entities in order to carry out the petroleum operations.

401. Regarding Farmee's obligations, as stated above, they derive, on the one hand, from the petroleum agreement concluded by the Farmor with NAMR, and on the other hand, from the contract and works agreed with Farmor. Obligations established in the petroleum agreement have a binding effect deriving from the law and may affect – in case of their non-fulfilment – the very existence of the contract. Most often, however, they attract penalties consisting in amounts to be paid to NAMR which will further deliver them to the state's budget. The amounts to be paid are the estimated amounts that were to be spent on operations notified to and accepted by NAMR, which have not been performed as agreed due to holders' fault. Regarding

476. For example, in some of the studied contracts although the parties have established an interest share of 50%–50%, respectively 35%, 14% and 51%, the Farmee assumed the obligation to pay 75% respectively 80% of total joint works costs until the discovery of deposits, respectively until the completion of the first phase of the joint works. As it can be easily observed, the contracting parties enjoy great freedom in determining their own rights and obligations.

the obligations deriving exclusively from Farm-out agreement, they must be fulfilled jointly, unless the Farmee has acquired a certain area within the perimeter,[477] under general legal consequences governing contractual liability.

III. Rights and Obligations of Third Parties: NAMR

402. The transfer of rights deriving from a petroleum agreement obviously has an impact on the initial concession, especially if the transfer is only partial. Regarding the rights of NAMR, given its grantor status, the Agency is entitled to maintain its rights established by the petroleum agreement. As shown earlier, these rights may be subject to renegotiation, which could lead to a novation of the original contract. But in the absence of renegotiation and of NAMR's express consent thereof, the parties involved in a farm-out agreement cannot unilaterally modify NAMR rights.

403. As such, regardless of the number of transfers, NAMR is entitled *inter alia*: to follow the payment of royalties set, to have ownership of all raw data coming from petroleum operations (including geological, geophysical, petro-physical and engineering data), to suspend the contract (if the holder or holders do not respect a court decision on disputes arising in connection to the execution of petroleum operations, is/are subject to judicial reorganization or bankruptcy, is/are jeopardizing the performance petroleum operations or is/are causing serious violations of health and safety regulations), to terminate contract (in case of non-fulfilment of contractual obligations, in case of knowingly providing false information, in case of non-performance of the minimum works programme agreed, or in case the holder(s) perform petroleum operations without permits required by law).

404. When analysing the transfer request, NAMR must verify the fulfilment of the holder's due obligations assumed in the petroleum agreement. The law states that approval of the transfer cannot be given if the Agency finds that they are not met, unless the Farmee has expressly accepted to take them over. Thus, besides the obligation of verification, NAMR is endowed with the right to refuse the transfer, on the one hand, and with the allocation of contractual obligations, both present and future, on the other hand.

405. It would be interesting to consider whether the liability of the parties, in the case of a partial transfer of rights arising from a petroleum agreement, is either jointly or severely. It is obvious that if the Farmee accepted to assume the unperformed obligations of the Farmor, he will be held sole liable for their fulfilment (having been a novation by change of debtor), following that the Farmor will indemnify him later.[478] Regarding the present or subsequent obligations, the holders are each liable to the extent of the interest owned. Even if one of the owners is nominated as Operator, and is authorized to act on behalf of the other non-operator parties, it does not mean that the holders are held jointly liable, if there is no express

477. Article 34^2 of Petroleum Law.
478. *See* Art. 1432 para. 1 of NCC.

provision in this regard.[479] Most times though, NAMR, by the approval order of the transfer, expressly mentions that the parties will be held jointly liable. In such conditions, NAMR can hold any part of the contract as liable, and afterwards the situation will settle among the owners, according to the understanding existing between them.

IV. Rights and Obligations of Other Third Parties

406. By 'other parties' the monograph nominates here other holders of rights deriving from petroleum concession agreement by farm-out agreements previously concluded. Depending on the Joint Operation Agreement (JOA), these Farmees may or may not be Operators, which will affect to some extent the nature of their rights. But since the Operator's rights deriving from the JOA are not covered by this monograph, there will not be any further mentioning of them.

407. The rights of previous Farmees cannot be violated, although their existence can give rise to discussions or even litigation. But although they cannot be broken by future agreements, they may be affected by subsequent actions, especially the actions of the Farmee who was designated Operator. That is because their rights are being closely related to maintaining the petroleum agreement, which is under the control of the Operator who, in turn, can accidentally or intentionally terminate it. Accidental termination occurs where the Farmee who was designated Operator, fails, due to his negligence, to perform any the contractual obligations sanctioned with termination of the agreement. This gives rise to a right of compensation of the non-operator Farmees against the Operator Farmee. The nature and extent of compensation is normally defined by the JOA.

§4. Termination of Farm-Out Agreements

408. The farm-out agreement may be terminated for multiple causes and given that its fate is tied to the existence of the oil concession agreement virtually any failure of the latter will obviously affect the farm-out agreement as well. The contract may be terminated: (a) by reaching its term, (b) by expiry or termination of the concession (either voluntarily or involuntarily as shown above), (c) by giving up the concession or a part of it which is deemed essential by the holders, (d) by depleting the deposit, (e) by cancellation, in case of non-performance of either the obligations of the parties toward each other, or the obligations owed to the state, (f) by disappearance of one of the holders, (g) under any other contractual provision agreed by the parties.

479. According to Art. 1426 para. 2 of NCC in the absence of contrary stipulation, creditors and debtors of indivisible obligations are not jointly liable.

Chapter 5. Exploitation

§1. Experimental Exploitation

409. Experimental exploitation is the petroleum operation which concludes the exploration phase of a discovery and is performed[480] in order to determine the technical and economic conditions for capitalization, in order to substantiate the decision of going to the development and exploitation phase of the deposit[481] or, as the case may be, to withdraw, go into preservation or abandonment of wells as well as for gathering the necessary data for evaluating the geological resources and reserves, for projecting and performing the development and exploitation works.[482]

410. In order to conduct an experimental exploitation programme, the license holder must draft and submit to NAMR a request and a programme of works for approval.[483] The minimum content of the programme is established by NAMR.[484] The duration for which experimental works can be solicited is of a year, with the possibility of extension[485] for an additional six months.[486]

411. The law established several cases in which holders of petroleum agreement may solicit approval for experimental exploitation: new objectives revealed by exploration-evaluation rigs or by operations of withdrawal in exploitation rigs; objectives with possible or probable reserves[487] whose technical-economic conditions for exploitation have not yet been set; objectives whose industrial productivity has been priory established but for which no experimental exploitation has been conducted due to technical or economic reasons; and objectives for which conventional or unconventional methods of secondary and experimental recovery will be conducted.[488]

412. The termination of experimental exploitation must be notified to NAMR within thirty days from the date of termination, through a final report[489] containing the main conclusions drawn from the experimental exploration of the rig. On the

480. Experimental exploitation is a petroleum operation performed by the holders of petroleum agreements on their perimeters. Art. 2, Annex 1 of NAMR's President Order No. 1/2006 for approval of Technical instructions concerning the amendment of the rigs' regime and the approval of petroleum operations of experimental exploitation (hereinafter Order 1/2006).
481. The data to be obtained is described in more detail in Art. 3.1 of Annex 1 of Order 1/2006.
482. Article 1, Annex 1 of Order 1/2006.
483. Article 3.4, Annex 1 of Order 1/2006. The approval procedure is detailed in Arts 3.6–3.9, Annex 1 of Order 1/2006.
484. Article 3.5, Annex 1 of Order 1/2006.
485. The extension requires NAMr's approval and it is granted by request on the basis of a memo containing the results of the petroleum operations already conducted together with the reasons for which the extension is requested and the supplementary data that is expected to be obtained in the solicited period.
486. Article 4.1, Annex 1 of Order 1/2006.
487. For proven reserves the law does not allow experimental exploitation operations. Art. 3.3, Annex 1 of Order 1/2006.
488. Article 3.2, Annex 1 of Order 1/2006.
489. The content of the report is set in Art. 5.1, Annex 1 to Order 1/2006.

basis of the report, the interested party shall solicit, as the case may be, moving the rig into final exploitation, withdrawing to a different level, switching the rig to conservation or its abandonment. Within thirty days from the report's submission, NAMR shall analyse the report and approve, ask for its amendment or decline the proposed solution.[490]

413. Where the exploration-evaluation rig has revealed a deposit which is deemed to have 'proven' reserves, the rig shall be shifted to final exploitation by notifying NAMR.

§2. The Annual Programme

414. According to Article 48 of Petroleum Law, corroborated with Article 88, letter b) of the Norms and Article 2 of NAMR's President Order 194/2006 concerning the approval of the Instructions regarding the drafting of the annual programme for petroleum exploitation (hereinafter Order 194/2006) all holders of petroleum agreements must draft an annual programme which they must submit for approval[491] to the competent authorities.

415. The annual programme will contain and detail the provisions concerning the production and the works established through the technical-economic studies approved by NAMR for the year in course, as well as the petroleum operations necessary for the performance of the exploitation process, which were not caught in the technical-economic studies, but are justified by the evolution of the exploitation process.[492]

416. The content of the programme is left up to the holders of the petroleum agreements, however, the law requires a minimum amount of information: (a) a technical memo[493] concerning the fulfilment of the previous year's programme together with a justification of any modifications that occur from the approved schedule, (b) annexes concerning the volume and the staging of petroleum operations, (c) the programme of technical, organizational, economic and protective measures concerning the deposit and the environment and (d) technical measures for conserving the deposit and restoring the environment in case of abandoning the concession.[494]

490. Article 3.13–Article 3.15, Annex 1 of Order 1/2006.
491. The approval procedure is detailed in s. 3, Arts 12–18 of Order 194/2006.
492. Article 4 of Order 194/2006.
493. The data to be contained in the technical memo is detailed in Art. 11 of Order 194/2006.
494. Article 10 of Order 194/2006.

Chapter 6. Transportation

417. According to Romanian law the national transportation system of oil is the property of the Romanian state and it is deemed an objective of strategic importance. The right to operate the national transportation system of oil is granted by concession to a holder of a petroleum agreement. Any investments performed by the holder throughout the concession's term, from his own resources, concerning the development and modernization of the national transportation system, are also public property of the state. However, for accounting reasons, the aforementioned investments are considered corporeal assets subjected to depreciation.[495]

418. The license holder has the obligation to allow access to the pipelines, notwithstanding whether they are part of the national transportation system (meaning public property) or not (meaning private property) under the conditions of the law.[496]

§1. The Transit of Oil

419. The transit of oil takes place through main pipelines, on contractual basis, with full observance of applicable national legislation and international treaties ratified by Romania.

420. The contracts for transit are negotiated between Romanian designated entities and similar entities designated by other involved states, on the basis of the covenants and other contractual terms established through the mandate given by the competent authorities. The terms and conditions established by the transit contracts will not include unjustified restrictions or any other clauses that would affect the security of delivery or the quality of the services, taking into consideration the production reserve, the storage capacity and the efficiency of the operating system in place.[497]

§2. Transportation of Oil through the National Transportation System

421. According to a report from 2014[498] the national transportation system through pipelines has a total length of 3,809 km out of which only 3,161 were used for the transportation of crude oil, gasoline, condensate and ethane. However, the

495. Article 14 of Petroleum Law.
496. Article 15, para. 2 of Petroleum Law.
497. Articles 16 and 17 of Petroleum Law.
498. Raport trimestrial privind activitatea economico-financiara a SC Conpet SA la data si pentru perioada de noua luni incheiata la 30 septembrie 2014, available on line, in Romanian, at http://www.conpet.ro/Actionariat%5CDocument%5CRaportPeriodic%5C2014%5CCOTE%2005.11.2014%20-%20Raport%20Trimestrial%20Trim.%20III%202014%20.pdf, last visited 23 Mar. 2018, page 2 of 20.

usability level of the national system dropped from 28.8% in 2009 to 27.12% due to the fact that one of the major oil producers has decreased the transported quantity after building its own maritime terminal.

422. The transportation of oil through the National Transportation System is defined by law as a public service of national interest and strategic importance.[499] As a direct consequence, the concessionaire of the National Transportation System is deemed a common transporter and must ensure free, transparent and non-discriminatory 'access' to the 'available capacity' of the system to all applicants.[500]

423. By 'available capacity' the law understands the difference between the total physical capacity of the system and the capacity undertaken by the transporter through contracts concluded and into force at the date of the applicant's requests, while assuring at the same time: (a) the transport of the oil belonging to the transporter, (b) the fulfilment of the common transporter's public service obligations, (c) the maintaining of a reserve capacity for safe, flexible and efficient operation of the system, according to the technical norms issued by the transporter and approved by the NAMR.[501]

424. By 'access' to the available capacity of the national transportation system, the law understands the right of any authorized applicant to benefit from effective connection to the system and/or to transport his oil or natural gas, according to the terms of the transportation contract in exchange for payment of the transportation tariffs.[502]

I. The Common Transporter

425. The law did not define directly who can act as a common transporter, but it specifically stated that legal persons which are controlled in any way by a third legal person which is directly or indirectly engaged in activities of production, refining or commercializing oil or which are simultaneously controlled by natural or legal persons which have control over the legal persons already mentioned, may not be common transporters.[503]

426. The common transporter has a set of positive and negative obligations. As positive obligations the transporter needs to ensure the qualitative and quantitative integrity of the transported oil and to ensure a reserve of capacity in order to maintain the integrity of the system. As negative obligations the law mentions that the legal person acting as common transporter should not engage, directly or indirectly,

499. Article 19 of Petroleum Law.
500. Article 20, para. 1 of Petroleum Law.
501. Article 20, para. 2 of Petroleum Law.
502. Article 20, para. 3 of Petroleum Law.
503. Article 23, para. 1 of Petroleum Law. However, Art. 24 allows the common transporter to sell and purchase oil under the conditions and limits established by NAMR for the sole reason of ensuring the physical balance and maintaining the operational parameters of the national transport system.

in any petroleum operations of exploration, development and exploitation and any processing, refining or commercializing activities.[504]

427. The law established several exemptions from the obligation of the transporter to ensure access to the national transportation system. These exemptions cover situations where: (a) there is no available capacity in the main pipelines; (b) there are technical, operations or safety reasons which could affect the security of the system, (c) the quality of the oil does not abide to the established standards. However, any refusal to grant access to the system must be motivated in writing and communicated to the applicant and NAMR.[505] The competent authority is empowered to investigate, analyse and resolve on its own initiative any case in which the common transporter refused to grant access to the system.[506]

II. The Framework Agreement for Transportation of Crude Oil, Gasoline, Condensate and Liquid Ethane

428. The contract for transport of crude oil, gasoline, condensate and liquid ethane is based on the Framework Agreement approved by NAMR.[507] This leaves to the signatory parties a little room for manoeuvre, given the fact that transportation tariffs are also established by law (*See infra* section IV.D.). Basically parties may negotiate only on issues such as: duration, method of payment, damages.

III. Safety and Protection Areas

429. In order to ensure the maintenance of the national transportation system and to eliminate possible dangers that would occur as a result of the system's exploitation, the law establishes safety and protection areas.

430. The safety and protection areas are determined in full observance of the legal and technical instructions issued and approved by NAMR.[508] Currently minimum safety distances are established through Annex 1 of Order 196/2006, and depend on the type of industrial, commercial, administrative or social objectives located in its proximity.[509] Protection areas, on the other hand, are established at 5 m on each side of the pipelines. Additionally, the law prohibits: any construction works (even temporary ones) in the protection area, the obstruction or limitation of

504. Article 21, para. 3 corroborated with Art. 23, para. 2 of Petroleum Law.
505. Article 21, paras 1–2 of Petroleum Law.
506. Article 22 of Petroleum Law.
507. *See* NAMR's President Order No. 247/2011 concerning the approval of the Framework agreement for transportation of crude oil, gasoline, condensate and liquid ethane for 2012 (hereinafter Order 247/2011).
508. NAMR's President Order No. 196/2006 concerning the approval of technical standards and regulations for safety and protection areas related to the National Transportation System of crude oil, gasoline, condensate and ethane (hereinafter Order 196/2006).
509. Articles 2 and 3 of Order 196/2006.

access and circulation in the protection area as well as installation of any services or utilities that do not serve the pipelines concerned.[510]

431. Where the safety areas are located on properties belonging to third parties a legal encumbrance is established,[511] according to civil law,[512] and on the basis of the principle of the smallest interference with the property right.[513] The right to access the safety area is negotiated between the license holders and the land owners[514] and it is exercised in exchange of an annual royalty towards the owners of the affected lands for the entire duration of the encumbrance.[515] The duration of the legal encumbrance is equal to the duration of the petroleum operations.[516]

432. On the one hand, the transporters have the right to perform interventions only with the prior notification of the land's owner.[517] However, on the other hand, they have a right to access the safety and protection areas permanently, occasionally or urgently, depending on the nature of the operations to be performed.[518] Any compensations for damages caused by the occasional or the urgent works performed are to be established through direct negotiation with the aggrieved party, on the basis of an expertise.[519] Only where the parties fail to reach an agreement, the amount shall be established by the court.[520]

IV. Transportation Tariffs

433. Transportation tariffs are established annually[521] by the competent authority, NAMR[522] on the basis of their own methodology.[523] The transport tariffs represent the equivalent of the transport services performed by the holder of the concession, as common transporter, while the services covered designate all the activities and operations performed by the holder of the transport agreement for or related to the access and usage of the system.[524] The tariffs are established differently for the two subsystems which form the National System of Transportation: on

---

510. Articles 12–13 of Order 196/2006.
511. Article 25, para. 1 of Petroleum Law.
512. Article 755 and the following of NCC.
513. Article 23 of Order 196/2006.
514. Article 25 of Order 196/2006.
515. Article 24, para. 1 of Order 196/2006. According to para. 3, the level of the annual royalty is established by legal expertise.
516. Article 22 of Order 196/2006.
517. Article 25, para. 4 of Petroleum Law.
518. Article 17 of Order 196/2006.
519. Article 25, para. 4 of Petroleum Law corroborated with Art. 19 of Order 196/2006.
520. Article 25, para. 4 of Petroleum Law corroborated with Art. 26 of Order 196/2006.
521. Article 2, para. 5 of Annex 1 to NAMR's President Order No. 53/2008 for approving the Instructions concerning the criteria, methodology and procedure for establishing the regulated tariffs for transport of crude oil, gasoline, condensate and ethane (hereinafter Order 53/2008).
522. Article 20, para. 4 of Petroleum Law.
523. Article 1, para. 1 of Annex 1 to Order 53/2008. The detailed methodology of calculating the regulated transportation tariffs is contained in Chapter 2, Art. 3 of Annex 1 to Order 53/2008.
524. Article 2, paras 1 and 2 of Annex 1 to Order 53/2008.

the one hand, the subsystem designed for oil, gasoline, condensate and ethane from internal production and, on the other hand, those coming from imports,[525] on the basis of the proposal received from concessionaire of the National Transportation System and after analysis and approval by NAMR.[526]

434. As of 3 February 2015, the tariffs are established through NAMR's President Order No. 12/2015 concerning the amendment of NAMR's President Order No. 13/2010, concerning the approval of transportation tariffs through the National Transportation System of crude oil, gasoline, condensate and ethane.

525. Article 2, para. 3 of Annex 1 to Order 53/2008.
526. Article 4, para. 1 of Annex 1 to Order 53/2008.

Chapter 7. Distribution and Consumption

435. In 2007 the Romanian petroleum market was estimated at EUR 9 billion, while in 2011 it was already above EUR 10 billion. At the same time, the consumption in 2007 was around 3 million tons of petrol and 4 million tons of diesel fuel. Beside engine fuels, the Romanian industry also needed around 7.3 million tons of fuel. However, in 2010 the fuel consumption per capita was 290 litre/annum, much lower than the 550 l/annum recorded for Central Europe.

436. The distribution market is still concentrated in the hands of a few number of players. Basically, six companies own the ten existing refineries, five of them covering 85% of the refining capacity of the country. On the refining market, Petrom holds 52%, Rompetrol 29% and Lukoil 17%. The total refining capacity of these ten refineries is 34 million tons per year, much more above the national consumption of Romania. Since the capacity is bigger than the production, Romania exports lots of oil products such as bitumen, fertilizers, and lubricants.

437. The Romanian engine fuels sales (petrol and diesel) has varied in the last five to ten years, largely due to the financial crisis and increase in prices. For example, the price of petrol has increased with 93% in Romania, during 2009–2012, while the price of diesel fuel has increased with 68%, which had an impact on consumption. The prices for fuels contain two categories of costs. On the one hand, there are the VAT, excise and other taxes which oil companies must pay to the state, on the other hand, there are the costs of processing raw materials, trade margins and financial cost. In the end, the price paid by the consumer at the pump is divided between the state and the oil companies.

438. The petroleum market continues to be dominated by OMV Petrom, who is still the largest tax payer to the state's budget, with 10% of the Gross Domestic Product. However, if Romania fails to discover, explore and exploit new deposits, the unanimous opinion is that its energetic dependence towards other states will increase severely. It is estimated that petroleum imports will reach 40%–50% in 2030 due to an increase of raw material demand and natural decline of domestic production of hydrocarbons.

439. Despite all the above, the domestic petroleum production has increased by 4% in 2013, Romania maintaining its position among the top ten petroleum producers in Europe and Eurasia. However, the consumption recorded a decrease of 1.7% for the same year.[527] 2013 has also recorded a decrease of 10% in oil imports compared to the previous year. Nevertheless, in 2017, the demand for fuels increased by 6.7% due to Romania's economic growth and large number of cars, combined with

527. http://www.bursa.ro/bp-productia-de-petrol-a-romaniei-a-crescut-cu-4-procente-in-2013-239753&s=companii_afaceri&articol=239753.html, last visited on 23 Mar. 2018.

a decrease of 4% in the domestic production. As a result, official figure from the National Institute for Statistics showed in 2017 a 3% increase in oil imports compared to 2016. In 2017, Romania imported 7.75 million tons of oil, while domestic production generated only 3.42 million tons oil.[528]

528. *See* Ergynomics – Consumul de carburanti creste, Romania importa tot mai mult titei, 25 Feb. 2018, available only in Romanian, http://www.energynomics.ro/ro/consumul-de-carburanti-creste-romania-importa-tot-mai-mult-titei/, last visited 23 Mar. 2018.

Part IV. Regulatory Framework Concerning Coal

Chapter 1. General Introduction

440. The legal framework concerning coal is provided by Mines' Law No. 85/2003[529] with all the further amendments together with the Orders and Methodologies issued by the NAMR's president. The law regulates the mining activities[530] taking place in Romania and strives to stimulate the capitalization of mineral resources which are property of the state.[531] Other principles applied by the law refer to maximum *transparency* of mining activities, *fair competition*, *non-discrimination* between types of property, capital origin and operators' nationality. Furthermore, the Romanian law encourages investments in the mining field through administrative and tax incentives, by guaranteeing investors free recovery of their investment and free usage of their profits.

441. The provisions of Mines' Law cover a wide range of mineral resources such as: coal, ferrous and non-ferrous minerals, aluminium, noble metals, radioactive, rare and dispersed haloid salts, non-metallic useful substances, useful rocks, precious and semiprecious stones, peat, mud and therapeutic peat, bituminous rocks, non-combustible gas, geothermal waters and the gas accompanying them, mineral water (carbonated and still), therapeutic mineral waters and mining residues from dumps and tailings dams, groundwater for drinking and industrial activities of prospecting, exploration, appraisal and accounting reserves.[532]

529. Mines' Law No. 85/2003 was published in the Official Gazette of Romania, Part I, No. 197/27.03.2003 and was subjected to numerous amendments in 2004, 2005, 2007, 2009, 2011, 2012, 2013, 2015, and 2016 (hereinafter Mines' Law).
530. According to Art. 3, point 1 of Mines' Law, mining activities are defined as 'the set of works concerning the exploration, development, exploitation, preparation/processing, concentration, commercialization of mining products, conservation and closure of mines, including restoration and rehabilitation of the environment'. All these activities are defined in detail in Art. 3 of Mines' Law.
531. Article 1 of Mines' Law reiterates the Constitutional provision according to which all mineral resources located on the territory, in the subsoil of the country or the continental shelf within the continental shelf of the Black sea's economic area of Romania constitute the exclusive object of public property, and belong to the Romanian state.
532. Article 2 of Mines' Law and Application Guidelines to Mines' Law from 14 Oct. 2003 (hereinafter the Guidelines to Mines' Law). The Guidelines to Mines' Law were published in the Official Gazette of Romania, Part I, No. 772/04.11.2003.

442. The competent authority representing the state's interest with respect to mineral resources is the NAMR.[533]

443. Mining activities may be performed by natural[534] and legal persons which are registered, according to the law, and are specialized and authorized to undertake such activities or they organize themselves for this purpose. The economic and technical documentation as well as the expertise related to the mining activities are prepared by natural persons authorized thereof by the competent authority. Mining activities are performed solely within the boundaries of perimeters authorized by NAMR, with the exception of activities performed by natural persons for personal usage under Article 29 of Mines' Law, where the authorization of the perimeter is not necessary.[535]

§1. General Regime for Coal

444. By law the perimeters for prospections, exploration or exploitation of coal are set through giving into administration or concession agreements, on the basis of a documentation whose content is established by NAMR's instructions. The provisions and covenants of the administration or concession agreements are also established by NAMR and are to be negotiated with the concessionaire or the administrator.[536] While concession agreements may concluded with either national or foreign legal persons, giving into administration can only be granted to public institutions.[537]

445. In case the right for performing mining operations is acquired by a foreign legal person, then within ninety days from the entering into force of the license, the

533. Article 3, point 4 corroborated with Art 54 of Mines' Law, as amended by GEO No. 40/2016 concerning the establishment of certain measures at central public administration level and for the modification and completion of certain normative acts, published in the Official Gazette of Romania, Part I, No. 490/30.06.2016 (hereinafter GEO No. 40/2016).

534. Natural persons are allowed to perform mining activities in certain situations described in Arts 28–30 of Mines' Law. Thus, a natural person may extract rocks for constructions and accumulations of peat in determined quantities, for maximum one year and on the basis of a permit issued by the competent authority. Where natural persons wish to exploit sands and gravels they must first obtain an exploitation permit which is issued after the natural persons have obtained all the approvals concerning the field of water management. Additionally, natural persons seeking permit for exploitation must post a bond for environmental restoration and pay the taxes and royalties for the activities performed (Art. 28). Rocks which can be used in constructions, peat and mineral waters which are located on private properties belonging to natural persons may be used by the owners only if they do not constitute already the object of a concession agreement and only for personal usage, excluding any commercialization. Personal usage is exempted from payment of taxes and royalties (Art. 29). Similarly, recovering gold from silts, outside a concession framework, is allowed for natural persons and legal persons alike on the basis of an exploitation permit issued by the competent authority, with payment of taxes and royalties according to the law (Art. 30).

535. Article 4 of Mines' Law.

536. Articles 23 and 24, para. 2 of the Guidelines to Mines' Law.

537. Article 13 of Mines' Law.

legal person shall set up and maintain a Romanian subsidiary for the whole duration of the license.[538] Failure to comply with this obligation may result in a fine in the amount of RON 30 to RON 60 million.[539]

446. Depending on the mining operations to be performed, the documents for giving into administration or concession[540] are:

- The permit for prospection.
- The license for giving into administration for exploration.
- The license of concession for exploration.
- The license for giving into administration for exploitation.
- The license of concession for exploitation.
- The permit for exploitation.
- The NAMR's approval regarding the association of the license holder with other legal persons in order to perform the mining activities that constitute the object of the concession or the administration.
- The NAMR's approval regarding the transfer of a license of concession for exploration or exploitation to another legal person.

447. The provisions of the licenses remain unchanged for their entire validity period.[541]

§2. Acquiring Rights of Usage and Access to the Lands on which Mining Activities are Performed

448. By law the right to use the lands necessary to perform mining activities within the exploration or exploitation perimeter can be acquired in several ways: by purchasing the land and the constructions situated on it (if the case) at the price established by the parties; by exchanging lands, together with the relocation of the aggrieved owner and the reconstruction of the buildings on the new land, at the expense of the beneficiary of the liberated land, as agreed by the parties; by renting the land for a determined period of time; by expropriating the land on grounds of public utility, with full observance of the law; by concession of the lands; by association between the land owner and the concession holder; by other means established by law.[542]

449. The law established several exemptions from the performance of mining activities. Hence, it is forbidden to perform mining activities on lands occupied by:

538. Article 23 of Mines' Law.
539. Article 57, para. 1 of Mines' Law.
540. Article 24, para. 1 of the Guidelines to Mines' Law.
541. Article 60, para. 1 of Mines' Law.
542. Article 6 of Mines' Law.

historical, cultural, or religious monuments; archaeological sites of great importance; natural reservations; sanitary protection areas and perimeters for hydrogeological protection of water supplies. The exempted lands are established by Government decision with the approval of competent authorities and with the establishment of damages and other compensatory measures.[543]

450. In order to access[544] the exploration/exploitation perimeters or to perform any activities related to mining the law established a legal easement right to cross over the adjacent lands in favour of permit holders. However, the determination of the lands that are to be affected must be done on the basis of the principle of minimum interference with the property right. As usually, exercising the easement right requires the payment of an annual fee to the owner of the affected lands either on the basis of an agreement between parties or, where the parties cannot agree, on the basis of a court decision. The easement right lasts as long as the mining activities.[545]

451. By law, one cannot obtain an easement right to cross over lands which are also exempted from mining activities.[546]

§3. Data and Information Concerning Mineral Resources/Reserves

452. All data and information concerning Romanian mineral resources belong to the Romanian state[547] and must be made available to the competent authority[548] who would handle their recording and management[549] by constituting a National Geologic Fund (NGF).[550] The holders of licenses and permits for mining activities are entitled to use the data for their own personal interest for the entire period of their mining activities. However, the competent authority, the holders of licenses and permits and other public bodies and institutions having competences in applying the Mines' Law are all bound by a duty of confidentiality[551] concerning the

543. Article 11 of Mines' Law.
544. According to Art. 8 of Mines' Law the access to the affected lands is established through negotiations between the holder of the mining activity and the owners with full observance of equality of treatment and equity principles. Any disputes between the parties are to be solved by courts, under emergency procedure.
545. Article 7 of Mines' Law.
546. Article 11, para. 1 of Mines' Law.
547. Article 5, para. 1 of Mines' Law corroborated with Art. 2 of Guidelines to Mines' Law.
548. The regime of reporting data and information concerning mineral resources and reserves is established in Arts 16–22 of Guidelines to Mines' Law. Art. 17 states clearly that all those holding a concession right concerning a mining activity or administrators of mining activities bear a duty to annually report to NAMR the records and the movement of mineral resources and reserves using a standard form and an explanatory note whose models are established by NAMR.
549. Article 5 para. 1 of Mines' Law corroborated with Art. 5 of Guidelines to Mines' Law.
550. Article 3 of Guidelines to Mines' Law.
551. Article 5, para. 4 of Mines' Law, corroborated with Art. 15 of Guidelines to Mines' Law. According to Art. 6 of Guidelines to Mines' Law the data and information constituting the NGF are classified and, according to Art. 9 of Guidelines to Mines' Law, their protection is an obligation of all involved institutions, economic agents or public authorities that hold them or issued them.

aforementioned data and information. Thus, sharing the data with a third party may take place only with the prior approval of NAMR, in accordance with the Guidelines to Mines' Law.[552]

§4. Termination of the Concession Agreement or of the Right to Administer

453. The concession agreements and the rights to administer are terminated: at the expiry date of the period for which they were granted; by unilateral termination of the license holder; by revocation of the license/permit by the competent authority; at the holder's request in case of events that would constitute force majeure and which determine the objective and definitive impossibility to fulfil some of its rights or obligations, as established by the license and which are essential for the performance of the mining activities; by exhausting the exploitable reserves (only in the case of concessions or right to administer the exploiting mining activities).[553]

454. Where force majeure events occur, the holder must notify the competent authority within five days since their occurrence, by submitting the documents certifying force majeure within fifteen days since the events. The documents must be issued by the institution which is legally habilitated thereof. The termination of the concession or the administration will come into effect after thirty days from the date of the notification.[554] However, in the case where the competent authority notifies to the holder, within the thirty days period, its refusal to acknowledge the event alleged as force majeure, the license holder may address the administrative court with a request to annul the license for force majeure, without payment of compensation for termination. Where the court rendered a positive decision, the competent authority is obliged to issue the decision confirming the termination of the concession or administration.[555]

455. The license holder is entitled to unilaterally terminate the license or the permit if, at the date of the notification of termination to NAMR, it complies with the following obligations: provides the competent authority with the documentation regarding the performed activity and its results until the notification date; provides the competent authority with the amount representing the value of the unperformed works that were established through the license or permit and due at the time of notification;[556] provides the competent authority with the document attesting the performance of the restoration works of the environment which was deteriorated as a result of the mining activities; provides the competent authority with the amount necessary for the unperformed conservation/closing works, related to the mining

552. Article 5 of Mines' Law.
553. Article 32 of Mines' Law.
554. Article 36, para. 1 of Mines' Law.
555. Article 36, paras 2–4 corroborated with Art. 32, paras 4 and 5 of Mines' Law.
556. Article 120 of Guidelines to Mines' Law.

activities performed until the moment of termination and for the post-closing monitoring programme of the environmental factors, established in the closing plan. All amounts paid by the license or permit holder go to the state's budget. NAMR has sixty days to verify the cumulative fulfilment of the above-mentioned conditions and, in case there is no refusal from its side, the license or permit are deemed terminated notwithstanding whether the competent authority has made a ruling or not. As a result, the competent authority must issue the decision concerning the termination of administration or concession.

456. Where the competent authority refuses to issue the aforementioned decision the interested party is entitled to address the administrative court having jurisdiction, within thirty days from notification. The prior procedure is not mandatory, which represents a legal exception from the provisions of the Law No. 554/2004 concerning administrative litigation.[557]

457. The decision of the administrative court is final and irrevocable and must be enforced by the competent authority within fifteen days, under the sanction of a fine in the amount of LEI 5 million for each day of delay. The collected amounts go to the state's budget. Additionally, refusal to comply with the court's decision may also attract criminal liability, according to the law.[558] The fine was established through the initial text of the law, in 2003, before the Romanian currency was denominated and LEI 10,000 became RON 1 (new Leu). Obviously the Romanian legislator did not intend for such an amount for daily fines for delay and this is a mere loophole. However, since the law was amended several times, without considering the denomination, it may be that the amount was kept for its strong deterrent effect.

458. NAMR's decision concerning termination of the concession or the administration right granted through the license is published in the Official Gazette of Romania, Part I, with the specific mentioning of the date when it comes into effect. The decision is also registered into the Mining Book, by the competent authority. Within three months after termination, all technical annexes and outbuildings whose dismantling would endanger the safety of the works will become state's property, without any compensation and free of any encumbrances. Within the same maximum term, the competent authority will establish the measures which will be taken with respect to the exploration or exploitation perimeter and its related facilities.[559]

557. Article 7 of Law No. 554/2004 on administrative litigation established a mandatory prior procedure which requires the aggrieved party to first try and settle the dispute with the public authority or public institution before addressing the administrative court. If Article 7 is corroborated with the provisions of Art. 193 of the New Code for Civil Procedure (NCCP) any action lacking the mandatory prior procedure would be dismissed as inadmissible.

558. Article 32 of Mines' Law. *See also* the provisions of Art. 36, para. 4 of Mines' Law.

559. Article 37, paras 1–2 and 4 of Mines' Law corroborated with Art. 135 of Guidelines to Mines' Law.

459. Once passed into state ownership, the outbuildings and technical annexes may be either given into usage to the new license holders (in which case they shall be nominated into an annex of the license) or shall be maintained, in case the operations were only stopped temporarily.[560]

560. Article 137 of Guidelines to Mines' Law.

Chapter 2. Authorities and Competences

§1. Competences of the Relevant Minister

460. Mines' Law establishes also a series of competences for the Relevant Minister such as: elaborating strategies and policies in the area of mineral resources, which are submitted to the Government for approval, and participating in their implementation; evaluating and approving the budget for fulfilment of mineral resources production at national mining companies; elaborating programmes for implementing the strategies and policies or the budged mentioned above, including prospection programmes realized from budgetary funds;[561] allocating funds from the budget for: refurbishment and development, prospection programmes, as established in the annual geologic plan, technological research, protection and restoration of environment, maintenance or closing of mines or pits, social protection measure or any other activities in the field; ensuring the drawing up of studies on the basis of which it selects, together with the competent authority in the field of finances, the mining products resulting from exploitation which is carried out with funds from the state's budget; approving together with the competent authorities in the field of labour and social solidarity and in the field of environment protection, guidelines and norms for labour, respectively, environment protection, specific to mining activities; approving feasibility studies and development plans of national mining companies or concerning the temporary or permanent cessation of exploitation works; approving draft laws or Government decisions regarding mining activities.[562]

§2. Rights and Obligations of the Competent Authority

461. The competent authority for applying the provisions of Mines' Law is NAMR, organized as a public institution of national interest, with legal personality, under the supervision of the Romanian Government.[563]

462. The Mines' Law establishes a series of main competences enjoyed by NAMR, such as: managing the mineral resources and the national geological fund, which are state's property; establishing the provisions and covenants of licenses and permits, granting licenses and permits established by law and regulating the performance of mining activities by guidelines and technical norms; receiving, verifying and recording data and information concerning mineral resources and reserves; establishing tariffs for issuance of documents; supervising the production of mineral products in order to establish the related royalties; supervising the application of measures established for the protection of surface and underground, both

561. *See* Art. 141, para. 1 of Guidelines to Mines' Law according to which the relevant minister shall manage the performance of the national geologic programme and ensure its financing from budgetary sources.
562. For a full list of competences *see* Art. 56 of Mines' Law.
563. Article 54 corroborated with Art. 3, point 4 of Mines' Law.

throughout and after the completion of mining activities; supervising holder's compliance with the provisions of the license or permit as well as those of the guidelines and norms; approving the documentation for the performance or cessation of mining activities; setting perimeters for hydrological protection for underground natural mineral, geothermal or thermos-mineral waters; collaborating with territorial authorities having competences in managing waters, environmental or labour protection; suspending mining activities performed outside the boundaries of the granted perimeter or which cause unjustified loss of reserves or the degradation of the deposit, until the determining causes have been removed; elaborating guidelines and technical norms for the application of the Mines' Law, together with the interested ministries; elaborating and updating the Mining Book and the Extractive Cadaster; finding and notifying any breach of the legal provisions; and elaborating draft laws and government decisions concerning mining activities.[564] All legal competences of NAMR concerning mining operations are detailed in Decision No. 1419/2009 concerning the organization and functioning of NAMR.[565]

463. By law, NAMR has also competences to establish misdemeanours and apply the legal sanctions through its authorized personnel,[566] in accordance with the legal regime applicable to misdemeanours.[567]

464. Another set of obligations concerning NAMR are those who fall within the responsibility of the territorial inspection compartments for mineral resources. These compartments of NAMR keep records of all documents, data and information which lay at the basis of calculating the mining taxes and royalties, for each granted perimeter. On the basis of activity reports and expense accounts received from holders of mining concessions, they file centralized quarterly reports which are sent to NAMR within ten days. The centralized reports filed with NAMR by the territorial inspectorates shall be accompanied by a memo stating any established inadvertences and the measures and/or sanctions applied during the covered period.[568]

---

564. Article 55 of Mines' Law.
565. Decision No. 1419/2009 concerning the organization and functioning of the National Agency for Mineral Resources has been published in the Official Gazette of Romania, Part I, No. 842/07.12.2009 and has been amended by Decision No. 550/2013 and Decision No. 1062/2014.
566. Article 59, para. 1 of Mines' Law.
567. *See* Government Ordinance No. 2/2001 concerning the legal regime of misdemeanours, published in the Official Gazette of Romania, Part I, No. 410/25.07.2001 with subsequent amendments, the latest one dating from 2016.
568. Article D, points 1, 2, and 4 of Guidelines on Mines Law.

Chapter 3. Licenses and Permits: Right and Obligations. Other Duties

§1. Rights and Obligations of License or Permit Holders

465. The law provides the license or permit holders with a set of rights such as: right to access the necessary lands for carrying out the mining activities, within the boundaries of the perimeter; to perform all mining activities stipulated in the license or permit, within the boundaries of the perimeter; to dispose of all mining products made; to use, with the observance of all legal provisions concerning water management and environmental protection, the surface and underground water sources needed for the performance of mining activities; to associate with other legal persons in order to carry out the mining activities provided in the license, with the approval of the competent authority; to obtain from NAMR all data and information needed for the carrying out of the mining activities, to possess and use these data and information as well as those obtained from its own activities, for the whole validity period of the license or permit; to cease activity for a period longer than sixty days, with the approval of NAMR; and to build roads, bridges, railroads, electricity networks and any other infrastructure facilities required for the carrying out of mining activities, with full observance of applicable legal provisions.

466. The list of obligations is fairly longer and refers to several main categories: payment of revenues, obtaining and maintaining required authorizations, carrying out mining works, data gathering and environmental protection. The law expects the license or permit holder: to observe the provisions of the law, the guidelines and the license or permit; to elaborate, before the start and throughout the mining activities, technical and economic documentations for their performance, for obtaining the approval or the authorization for the environment according to the applicable legislation and submit them to the competent authority; to start mining activities in maximum 210 days since the enter into force of the license; to obtain, draw up, update and communicate to the competent authority, at the established deadlines, all data, information and documents provided for in the license or permit with respect to the performed mining activities as well as with respect to the obtained results, in order to be recorded in the Mining Book and the Extractive Cadaster; to inform the competent authority with respect to controls performed by local authorities having competences in environmental and labour protection; to keep confidentiality over data and information obtained from the competent authority and relevant ministry; to carry out and finalize the works required for environmental restoration; not to transport, use, handle or deposit during performed works radioactive waste, toxic or dangerous substances, or any other polluting substances, other than with full observance of the legal provisions thereof; to perform at the end of the concession all maintenance and/or closure of the mine, according to the closing plan, including post-closing monitoring of the environment;[569] to maintain for the whole period of the exploitation the financial bond for environmental restoration and to pay the taxes and royalties[570] related to the mining activities.[571]

467. Failure to observe most of the aforementioned obligations may result in payment of fines, varying between RON 30 and RON 500 million,[572] unless the breach does not amount to a criminal offence. For example, performing mining activities in the absence of a permit or license (with the exceptions provided by the law) attracts criminal liability which could amount to six months to three years imprisonment.[573] In order to increase the deterrent effect of its sanctions Romanian Mines' Law also provided that repeated breaches of the above-mentioned obligations are punished with the double of the maximum amount of the fine established by law for the breach in question.[574] Furthermore, repeated breaches of the holder concerning his legal obligations may result in withdrawal of the license or the permit.[575]

468. The law established a set of additional obligations towards the relevant minister for companies or national mining undertakings such as: to base and communicate annually, at established deadlines, the financing volumes needed for the carrying out of the license's works; to communicate the required information with respect to the undertaken investments and the resulted production as well as with respect to the use of funds provided by the state's budget; to perform, in case of privatization, maintenance or closing works of mines and pits as well as works for environmental restoration for the surfaces affected until the moment of privatization.[576]

469. In addition to all the above, the holders are liable in tort for all damages caused by their mining activities to other natural or legal persons, from their own

569. This obligation is also specified in Art. 37, para. 3 of Mines' Law which states that holders are liable both materially and financially for the restoration of all environmental factors affected by the mining activities, according to the environmental restoration plan approved by the competent authority.

570. In close connection with the obligation to pay royalties, Art. 46 of Mines' Law has established an obligation for license holders to transmit to the competent authority any relevant data and information required for calculating the mining royalty due. With respect to the obligation to pay taxes and royalties and the ancillary obligations such as registering with the fiscal authorities as payers of taxes and royalties deriving from mining operations or registering with the compartment for territorial inspection belonging to NAMR *see also* Article A of The NAMR's Technical Guideline Concerning Tracking, Calculating, and Paying of Taxes on Mining Activities and Mining Royalties of 28 Sept. 2009 (hereinafter Guidelines on Mining Taxes), published in the Official Gazette of Romania, Part I, No. 648/01.10.2009, as amended by Order No. 158/2011 and Order 81/2013.

571. For a full list of obligations *see* Art. 39 para. 1 of Mines' Law.

572. According to Art. 58, para. 4, the limits of the amounts shall be periodically amended with the inflation rate, by Government decision.

573. Article 57, paras 1 and 2 of Mines' Law.

574. Article 58, para. 1 of Mines' Law.

575. Article 58, para. 2 of Mines' Law.

576. Article 39, para. 2 of Mines' Law.

negligence, until the date of the expiry or unilateral termination of the license, notwithstanding whether such damages are discovered after the termination of the concession or administration.[577] In the case where the holders fail to pay on time the taxes and royalties related to the mining activities, they will also be liable for interest and delay damages established according to the fiscal legislation into force.[578]

470. In order to ensure rapid intervention in case of an emergency caused by an accident or malfunctioning, the holders must set up stations for mining salvation, in full accordance with Romanian legislation in the field of environment and labour protection. These stations are set up either in mines and pits or at central units. The set-up decision belongs to the holder who must obtain first the approval of the territorial inspectorate for labour protection and territorial inspectorate for environment.

§2. Transfer of a License for Concession

471. Rights and obligations acquired through a concession license may be transferred to a third party only in accordance with the provisions of Mines' Law as detailed in its Guidelines.[579] However, the transfer may take place solely with the prior written approval of NAMR. Any transfer lacking NAMR's approval is deemed to be automatically null and void.[580] Additionally, NAMR is entitled to grant to other legal persons the right to explore or exploit other mineral resources in the same perimeter, with the approval of the license holder.[581]

472. The legal procedure requires that the parties interested in the transfer jointly solicit it, in writing, through a document bearing the stamps and the signatures of their legal representatives, which is to be accompanied by several documents. These documents are: a certificate issued by the Trade Registry concerning the legal person to whom the license is transferred; a memorandum prepared by the license holder, concerning the reasons behind the request for transfer; the decisions of the management bodies of the two legal persons[582] which approve the transfer of the license; a certified copy of the constitutive act and the incorporation certificate of the legal person to which the license is being transferred; documents concerning the technical and financial capabilities of the person to which the license is being transferred; a report concerning the mining activity performed by the holder until

577. Article 37 para. 5 of Mines' Law.
578. Article 47, para. 3 of Mines' Law.
579. Article 26 of Mines' Law provides that rights acquired through concessions may only be transferred under the procedure established by law in Art. 24.
580. Article 24, para. 1 of Mines' Law corroborated with Art. 109 of Guidelines to Mines' Law.
581. Article 25 of Mines' Law.
582. From the wording of the Guidelines to Mines' Law it results that the transfer cannot take part between the license holder and multiple legal persons at the same time. However, similarly to what was detailed with respect to transfers concerning oil concessions, there is no legal provision banning the license holder to transfer only parts of his rights and obligations to multiple legal persons through separate deeds. Thus, the legal reference to 'two legal persons' must not be construed in any way to mean that the transfer can only be from a legal person to one another legal person.

the moment of the request for transfer, which expressly state the stage of completion of the undertaken obligations and the obtained results; and the approval of the transfer issued by minister which coordinates or under whose subordination lies any of the legal persons involved (where the case may be).[583]

473. However, in order to give its approval, NAMR verifies the cumulative fulfilment of the following criteria: the license for concession must be valid; the legal person to whom the license is being transferred must not have unpaid due taxes to the state's budgets;[584] the obligations undertaken by the license holder must be performed according to the schedule and the elapsed period until the request for transfer unless the person to whom the license is being transferred agrees to take them over; the person to whom the license is being transferred must prove both the technical and financial capacity to perform the undertaken obligations as established through the initial concession agreement; the transfer must not alter the initial terms of the concession agreement and, lastly, the person to whom the license is being transferred must be specialized and certified by NAMR to perform mining activities.[585]

474. It is obvious from the above that the same policy considerations behind state's control over the transfer of concession right and obligations concerning oil also apply with respect to coal or other mineral resources. The main interest of the state is that the concession agreement is carried out by authorized and specialized legal persons, having both the know-how and the financial power to comply with the undertaken obligations.

475. Once NAMR receives the required documents and verifies that the aforementioned criteria are met, it has sixty days to issue its approval for transfer. In case of a positive outcome, the transfer is approved by Order issued by NAMR's president.[586]

476. The Mines' Law also provides for cases where the transfer of the license is required for reasons related to changes in the Articles of Incorporation of the license holder in case of reorganization, mergers or acquisitions, or any other reasons.[587] In this regard, the successor entity must notify NAMR on the takeover operation and must expressly solicit the granting of the license. The request must be accompanied by documents attesting the takeover of the initial holder's patrimony and NAMR shall issue its decision within thirty days from the moment it received it.[588]

583. Article 111 of the Guidelines to Mines' Law.
584. Article 59, para. 1 of Guidelines to Mines' Law. According to Art. 60, para. 1, letter b) of the Guidelines to Mines' Law the payment of all taxes due to the state's budgets is proven by certificates issued thereof by the public authorities which administer those budgets.
585. Article 110 of Guidelines to Mines' Law.
586. Articles 112 and 113 of the Guidelines to Mines' Law.
587. Article 24, para. 3 of Mines' Law.
588. Article 114, paras 2 and 3 and Art. 115 of the Guidelines to Mines' Law.

477. In such cases, the license is granted to the legal successors of the initial holder, either on the basis of the contract between parties or on the basis of a judicial decision.[589] According to the Guidelines to Mines' Law, in case the license is granted by contract, the parties will conclude an addendum to the initial agreement, which will state the new entity which took over, in part or in total, of the initial holder's patrimony, including those deriving from the license. Thus, the novation regards only the legal entity which replaced the original holder and the designation of its legal representative. All other terms and conditions remain unchanged.[590]

478. However, the Mines' Law expressly bans the transfer of licenses for rights of administration.[591]

§3. Taxes, Royalties and Tariffs

479. As it was previously mentioned, one of the main obligations of license holders is to pay to the state's budget the annual taxes for prospection, exploration and exploitation of mineral resources together with a mining royalty.[592] However, payment of taxes and royalties does not exempt the license or permit holder from paying the other budgetary obligations due according to the fiscal legislation into force.[593]

480. The taxes[594] and royalties[595] to be paid are established by law. Currently, the royalty for coal is established at 4% of the value of the mining production.[596] The amount of taxes is updated each year, according to inflation rate. However, since taxes are calculated at the surface of the granted perimeters,[597] the amounts may be reduced in case the perimeters are reduced at the request of the holder.[598] Since royalties are expressed in euro, they are not subjected to amendments due to inflation, but the law specifies which currency exchange rate shall be used for calculating them.[599] Taxes are owed annually and are paid in advance for the next year

589. Article 24, para. 3 of Mines Law.
590. Article 114, para. 1 of the Guidelines to Mines' Law.
591. Article 24, para. 4 of Mines' Law.
592. Article 39 of Mines' Law corroborated with Article A, point 1 of Guidelines on Mining Taxes.
593. Article 49 of Mines' Law.
594. The current amount of taxes is established by Art. 44, paras 2–4 of Mines' Law, as amended by Decision No. 350/2015, concerning the update of the amount of taxes concerning mining activities covered by Mines' Law No 85/2003, published in the Official Gazette of Romania, Part I, No. 359/25.05.2015 (hereinafter Decision No. 350/2015).
595. The current amount of royalties is established by Art. 45, para. 1 of Mines' Law.
596. http://www.namr.ro/minning-law/mining-royalties-tariff/, last visited on 23 Mar. 2018.
597. Article B, point 1 of Guidelines on Mining Taxes.
598. Article 117, para. 3 of Guidelines to Mines' Law.
599. Article 45, para. 1^1 of Mines' Law states that royalties are to be paid at the official exchange rate RON/EUR established in the first working day of October of the previous year, as published in the Official Gazette of the European Union. Paragraph 2 of the same articles establishes that for natural mineral water, the royalty is established at the source, at a rate of EUR 4 per 1000 l, at the exchange rate of the Romanian National Bank valid at the date of payment.

until 31 December of the year in course[600] while royalties are owed from the day the production has started[601] and are owed quarterly, being due at the date of twenty-fifth of the first month of the next quarter.[602]

481. Additional terms for payment are detailed in the procedure for payment of taxes comprised in the Guidelines to Mines' Law[603] and the Guideline on Mining Taxes.[604] Thus, for the first year of activity, the holders are expected to pay the taxes at the date when the deed for concession or administration entered into force, while holders of permits for exploitation are expected to pay the taxes at the date when the permit is issued.[605]

482. In the case of licenses for exploration and exploitation, the financial bond for environmental restoration is established annually as well, in the first month of the covered period, while in the case of exploitation permits, the bond is due at issuance date. By law, the financial bond may not be smaller than the value of the restoration works for that year.[606]

483. The competent authority is entitled to establish[607] and perceive tariffs for any documents issued in the exercise of its powers. The extra budgetary revenues obtained from these tariffs are used for covering the expenses generated by the composition of data packages and the calls for public tenders for offers, organizing rounds for presenting the conditions for participating in the public offer, holding tenders and elaborating studies, expertise and technical consultancy as well as any other material expenses and supplies.[608]

484. Last but not least, according to the provisions of Mines' Law, all amounts originating for the fines imposed on license or permit holders for breach of their legal obligations are income to the state's budget.[609]

§4. SUSPENSION OF THE LICENSE OR THE PERMIT

485. The competent authority is entitled to suspend the license or the permit when it establishes that: the holder does not comply with a court decision concerning any litigation arising from the performance of mining activities; the holder is

600. Article 44, paras 5–6 of Mines' Law and Art. 177, para. 1, letter b) of the Guidelines to Mines' Law corroborated with Article C, point 1 of Guidelines on Mining Taxes.
601. Article 118 of Guidelines to Mines' Law. The way to determine the basis for calculating the royalty owed is described in Art. 119 of Guidelines to Mines' Law.
602. Article 45, para. 4 of Mines' Law.
603. Articles 121–134 of Guidelines to Mines' Law.
604. Articles B and C of Guidelines on Mining Taxes.
605. Article 117, para. 3 of Guidelines to Mines' Law.
606. Articles 132 and 133 of the Guidelines to Mines' Law.
607. Tariffs are established through the Order of the President of NAMR and published in the Official Gazette of Romania, Part I.
608. Article 48 of Mines' Law.
609. Article 59, para. 2 of the Mines' Law.

under insolvency or bankruptcy proceedings; the holder's approval regarding environmental or labour protection has been suspended; the holder endangers, through the way it performs its mining activities, the future exploitation of the deposit, or infringes the guidelines regarding the protection and safe exploitation of the deposits; the competent authorities that the holder grossly infringes the norms concerning health and safety in labour.[610]

486. The suspension of the license or permit, for any of the aforementioned reasons, comes into effect from the moment when the competent authority has informed the holder thereof and produces effects until the moment when the cause of the suspension has ceased. However, the law states that the maximum term for suspension is one year. In case the holder's non-compliance is longer than a year, the competent authority is entitled to annul the license or the permit.[611]

487. Where the holder is unsatisfied with NAMR's decision regarding the suspension, he is entitled to contest it in administrative courts, within thirty days from the moment he received it. At the holder's request, and hearing the institution's position, the court may temporarily suspend the effects of NAMR's decision until it reaches a final and definitive solution. Similarly to procedure established for the cases where the holder unilaterally terminates the license or the permit, the prior procedure required by Law No. 554/2004 concerning administrative litigation is not mandatory. Failure to enforce the court's decision may attract fines in the amount of RON 5 million and criminal liability according to the law.[612]

§5. Cancellation of the License or Permit

488. As already stated, one reason for the competent authority to annul the license or the permit is the holder's non-compliance for a period longer than a year with the legal requirements, after his license or permit has been suspended.

489. However, the law has established several other reasons which may attract the cancellation of the license or of the permit such as: the holder does not comply with the undertaken obligations concerning the authorization and the deadline for starting the mining activities; the holder ceases mining activities for a period longer than sixty days without the approval of the competent authority; the holder uses exploiting methods and technologies other than those agreed upon in the development plan, without the approval of the competent authority; the holder performs activities without the approval of the competent authority regarding the annual working programme (in case of exploration) or the annual exploitation programme; the holder's approval or authorization concerning environmental or labour protection has been annulled; the holder knowingly provides the competent authority with

610. Article 33, para. 1 of Mines' Law.
611. Article 33, para. 2 corroborated with Art. 34, letter h) of the Mines' Law.
612. Article 35 of Mines' Law.

false data and information with respect to its mining activities and breaches the confidentiality provisions established in the license; and the holder does not pay within six months since the due date the taxes or royalties owned to the state.[613]

490. Holders unsatisfied with NAMR's decision of cancellation may challenge it in administrative court in similar conditions with the procedure established for challenging the decisions of suspension.[614]

491. As in the case of termination, within three months after cancellation, all technical annexes and outbuildings whose dismantling would endanger the safety of the works will become state's property, without any compensation and free of any encumbrances.[615] However, holders of annulled exploitation licenses remain liable for the integrity of the outbuildings and technical annexes which pass into state ownership for three months.[616]

§6. Authorizations

492. By law, each public authority is entitled to issue its own procedure for granting approvals or authorizations regarding environmental or labour protection needed for carrying out mining activities.[617]

493. In this respect, NAMR bears the obligation to inform the county or local councils and the county prefectures in whose jurisdiction lay the granted perimeters, in writing, within ten days from the date of entry into force of the exploitation licenses, of the mining activities and the perimeters affected by these licenses. The aforementioned local authorities have ninety days to amend and update the existent spatial and general urban planning in order to allow the performance of all operations needed for the mining activities.[618]

494. With respect to authorizations and permits for constructions needed for mining works, in order to ease the administrative burden imposed on investors, the law states that they are to be issued for all constructions needed and not separately for each construction. However, the authorizations and permits for constructions may impose conditions regarding parameters and functional characteristics for the buildings or the equipment in question.[619] Compliance with these conditions is monitored by license holders and verified by competent authorities throughout the technological testing period, which cannot be longer than six months from completion. Where the competent authorities notice full compliance, they must issue the

613. Article 34, letters a)-g) of Mines' Law.
614. Article 35 of Mines' Law and *supra* 425.
615. Article 37, paras 1–2 and 4 of Mines' Law corroborated with Art. 135 of Guidelines to Mines' Law.
616. Article 136 of Guidelines to Mines' Law.
617. Article 40 of Mines' Law.
618. Article 41 of Mines' Law.
619. Article 42 of Mines' Law.

authorization, approval or permit within ten days from the completion of the testing period, unless the law provides otherwise.[620]

495. Authorizations, approvals and all other permits necessary for carrying out the mining activities are valid for the entire validity period of the license or permits which lay at the basis of the said works. However, in case the conditions considered at the issuance of the authorizations, approvals and all other permits suffer any changes, then the holders of mining activities must request new ones.[621]

§7. Dispute Settlement and Arbitral Clauses

496. As seen above, the general approach of Romanian law with respect to dispute settlement concerning the interpretation and performance of licenses or permits is that it belongs to Romanian administrative courts.

497. However, the law has established an exception from the general rule. Thus, it is stated that Romanian courts shall have jurisdiction over disputes concerning interpretation and performance of licenses and permits only if the parties have not agreed for an arbitral court, notwithstanding whether it is a national or an international tribunal.[622]

620. Article 43, paras 3 and 4 of Mines' Law.
621. Article 43, para. 1 of Mines' Law.
622. Article 61 of Mines' Law.

Chapter 4. Production

§1. The Prospection Permit

498. The prospection permit enables the holder to perform evaluation and interpretation studies of the pre-existent data as well as works of geological mapping, drillings, surface excavations, laboratory tests, and any other works for establishing the general conditions favourable to accumulation of mineral resources, including coal.[623] The permit covers a determined perimeter where all prospection works can be performed. The permit is non-exclusive and it is issued by NAMR at the written request of interested national or foreign legal persons.[624]

499. The permit is issued for a period of maximum three years, which cannot be extended. The corresponding royalties must be paid in advance, each year. The holders of the permits bear the obligation to perform a minimum amount of works[625] which is established after negotiations with NAMR. In addition, the holders must submit quarterly and annual reports regarding the performed works and their value, supported by documents. Last but not least, within sixty days from the expiry of the permit, the holder must submit a final report containing all methods of investigation used, performed works, their value and the obtained results.[626]

500. If the holder of a permit for prospection wishes to obtain a license for exploration, then he must solicit[627] a public tender for offers and submit, together with the request, a report containing information with respect to the conducted investigations, performed works, the justification of the works' value and the retrieved results, solely for the perimeter for which he requests the license.[628] By law, holders of a prospection permit which participate in a public tender for offers concerning the perimeters where they have conducted their prospection will benefit from an additional scoring,[629] based on the criteria established by NAMR.[630] However, those holders who either fail to request a license for exploration or to participate in a public tender for offers concerning their perimeter (notwithstanding whether it is initiated by NAMR or other interested legal persons) will not benefit from the legal provisions concerning the additional scoring.[631]

---

623. Article 25 of the Guidelines to Mines' Law.
624. Article 14, para. 1 of Mines' Law corroborated with Art. 26 of the Guidelines to Mines' Law. The permit is issued in thirty days since the moment when the interested legal person has complied with all the requirements set by NAMR's technical guidelines.
625. The amount of works is usually correlated with the validity period of the permit and the surface of the prospection perimeter.
626. Article 14, paras 2–4 of Mines' Law.
627. The solicitation may be formulated either during the period of the prospection permit or within sixty days from the submission of the final report.
628. Article 27 of the Guidelines to Mines' Law.
629. Article 14, para. 5 of Mines' Law.
630. Article 15, para. 8 of Mines' Law.
631. Article 27, para. 3 of the Guidelines to Mines' Law.

§2. The License for Exploration

501. The license for exploration enables its holder to undertake specific studies and works necessary for the identification of the coal resources or reserves, for their qualitative and quantitative evaluation and for the determination of the technical and economic conditions for their capitalization. Basically, the purpose of exploration works is to fundament the decision regarding the capitalization of the deposit, to furnish all the data needed for planning and performing the exploitation works,[632] while maintaining a safe level of coal resources and reserves.[633]

502. Exploration works are performed on the basis of an exclusive license granted for any of the mineral resources discovered in the perimeter. The license is granted to the winner of a public tender for offers which is organized either at the request of interested national or foreign legal persons or at the initiative of the competent authority. However, the list containing perimeters for exploration is established by NAMR's Order, which is published in the Official Gazette of Romania, Part I.[634]

503. In order to participate at the public tender, national and foreign legal persons must submit their offer in the determined period of time, established by the competent authority through the Order approving the perimeters for exploration. By law, offers must contain the proposed exploration schedule, supporting documents concerning the technical and financial capacities of the offeror as well as other documents required by NAMR. The proposed schedule contains the annual exploration volume and the related expenses, both of them mandatory to be reached within the agreed timeframes.[635]

504. The terms of the public tender for offers, together with the selection criteria of the winning offer, the additional score granted to the holder of the prospection permit and any other aspects are detailed in the Guidelines to Mines' Law.[636]

505. The Guidelines contain provisions with respect to all stages of the public tender for offers and specifically refer to: the principles on which the public tender is based;[637] the publicity of the tender;[638] the required content of the offers;[639] the submission of the offers;[640] the Opening, Evaluation and Negotiation Committee[641]

632. Exploration works are defined in detail in Art. 28, para. 3 of the Guidelines to Mines' Law.
633. Article 28, paras 1–2 of Guidelines to Mines' Law.
634. Article 15, paras 1–4 of Mines' Law.
635. Article 15, paras 5–7 of Mines' Law.
636. Article 15, para. 8 of Mines' Law corroborated with Arts 33–90 of the Guidelines to Mines' Law.
637. Article 33 of the Guidelines to Mines' Law.
638. Article 41 of the Guidelines to Mines' Law.
639. Articles 44, 60, 63 and 65 of the Guidelines to Mines' Law.
640. Article 46 of the Guidelines to Mines' Law.
641. Articles 50, 52 and 53 of the Guidelines to Mines' Law.

and its duties;[642] the evaluation of the offers;[643] the opening of the offers;[644] the designation and communication of the winning offer;[645] the challenge of the tender's result;[646] the procedure for negotiating the programme for exploration and the project for rehabilitation of the environment.[647]

506. The license for exploration is valid for five years with the advance payment of yearly royalties[648] and will specify: its duration; the exploration perimeter; the works schedule, the measures for environmental protection and the rhythm for the performance of the agreed works, the required documentation and the rights and obligations of the holder, as provided by the law.[649]

507. The license is concluded in writing and enters into force upon the publication of the approving NAMR President's Order in the Official Gazette of Romania. The legal provisions governing the license at the time when it entered into force shall govern its legal regime throughout its entire existence, unless a more favourable regime for the holder has appeared.[650]

508. With respect to the works schedule the law states that it is mandatory and must be performed in full until the expiration of the license's period.[651] However, the Guidelines to Mines' Law provide that it may be amended during the performance of exploration works, within the confinements of the exploration works agreed upon within the license. In this regard, the holder of the license must file with NAMR a written request for approval. Even so, the physical works agreed for the first year of exploration cannot be diminished.

509. The perimeter's surface where the mining activities are performed is defined by topogeodezical coordinates and its shape and dimensions are corresponding to the geological structures and formations and the mineral resources which constitute the object of the exploration license. The holder is entitled to reduce the surface of the perimeter solely with the approval of the competent authority, on the basis of the stage's documents, with the obligation of performing in full all works agreed for the first exploration year. However, the holder must also provide evidence that all works required for restoring the environment on the relinquished surface have been performed.[652]

---

642. Article 54 of the Guidelines to Mines' Law.
643. Articles 59, 62 and 67, 68 and 70 of the Guidelines to Mines' Law.
644. Article 72 of the Guidelines to Mines' Law.
645. Articles 77 and 78 of the Guidelines to Mines' Law.
646. Article 80 of the Guidelines to Mines' Law.
647. Article 88 of the Guidelines to Mines' Law.
648. Article 16, para. 1 of Mines' Law.
649. Article 29 of the Guidelines to Mines' Law.
650. Article 21 of the Mines' Law.
651. Article 16, para. 2 of Mines' Law.
652. Article 16, paras 3 and 5 of Mines' Law.

510. The start of the mining activities established in the license is authorized in writing by the competent authority within 180 days since the entry into force of the license but no longer than 30 days after the holder has submitted cumulatively the following documents: the evidence of paying the mining activity tax, the evidence of posting the bond for environmental restoration, the certificate issued by the fiscal authorities, attesting that the holder does not have any outstanding obligations to the state, the environmental approval, and the approval of the competent authority for the annual works programme.[653]

511. License holders must submit quarterly and yearly reports to the competent authority with respect to the performed activities and their costs, at the established dates, as well as a final report concerning the applied methodology, the performed works, their respective costs and the obtained results. The final report must be submitted within sixty days of the license expiry date.[654]

512. Holders of exploration licenses must solicit, within ninety days from the submission of their final exploration report to NAMR, a license for exploitation,[655] otherwise they will be precluded from directly obtaining a license to exploit any of the resources discovered within the exploration perimeter.[656] Where such request has been made, the license holder is entitled to directly receive the license for exploitation for any of the discovered mineral resources. Additionally, he is entitled to continue, until the date of entry into force of the exploitation license, the performance of experimental exploitation, preservation and maintenance works of the mining activities undertaken in the exploration perimeter.[657]

653. Article 22 of Mines' Law.
654. Article 16, para. 4 of Mines' Law.
655. Article 29, para. 5 of the Guidelines to Mines' Law.
656. Article 17, para. 1 corroborated with Art. 18, para. 1, letter a) and Art. 20 of the Mines' Law.
657. Article 17 of Mines' Law.

Chapter 5. Exploitation

§1. LICENSE FOR EXPLOITATION

513. Mining exploitation can be performed only on the basis of an exclusive exploitation license which can be granted either directly to the exploration license holder or to the winner of a public tender for offers organized by the competent authority.[658] The initiative of concession of mining activities for exploitation belongs either to the NAMR or to the interested national or foreign legal person. As in the case of other concessions, the list containing the perimeters for exploitation is established by the competent authority and published in the Official Gazette of Romania. However, in order to comply with the technical requirements for protecting underground waters, only perimeters for which the prior approval of the competent authority for water management has been already obtained shall be offered for concession.[659] All persons interested in participating in the public tender must submit their offers[660] in a determined period of time set by NAMR's Order.[661]

514. The license is granted through negotiation on the basis of a request accompanied by a feasibility study concerning the capitalization of the mineral resources and the protection of the deposit, the plan to develop the exploitation,[662] drawn up according to the technical instructions issued by the competent authority, an impact study regarding the environment, a plan to restore the environment and a technical project, as set out by the technical instructions issued by the competent authority and an evaluation study concerning the social impact, according to the instructions issued by the competent authority in the area of labour and social solidarity together with the relevant minister. The maximum period of the license is twenty years, with the possibility for consecutive extensions of five years each. In exchange, the license holder bears the obligation to pay yearly a tax for the exploitation activities and a mining royalty as established by the law. In addition, the license holder must post a bond for environmental restoration.[663]

658. Article 18 of Mines' Law.
659. Article 19, para. 5 of Mines' Law corroborated with Art. 92 of Guidelines for Mines' Law.
660. Offers shall be accompanied with evidence concerning the technical and financial capacity of the applicant together with any other requested documents set by the competent authority through the procedure for the public tender. Article 19, para. 4 of the Mines' Law. The procedure for public tender is detailed in Arts 91–101 of the Guidelines to Mines' Law.
661. Article 19 of Mines' Law.
662. The content of the exploitation development plan is established in Art. 92, para. 2 of the Guidelines for Mines' Law and includes: the quantity of resources proposed to be exploited, the proposed duration of the concession, the expected production capacity for capitalization of the mineral resources/reserves, the period for reaching the maximum production capacity, the physical volumes and values for investments, the rate of recovery of resources/reserves, the resulting mining products, the volume of works for geological research for increasing the knowledge concerning the mineral resources/reserves, the physical volumes and values for environment recovery works, other parameters established and required by NAMR. According to para. 3, the proposals regarding the parameters for the exploitation development plan are minimal and mandatory for the winning bidder.
663. Article 20 of Mines' Law.

515. The license is concluded in writing and enters into force upon the publication of the approving Government's Decision in the Official Gazette of Romania. The legal provisions governing the license at the time when it entered into force shall govern its legal regime throughout its entire existence, unless a more favourable regime for the holder has appeared.[664]

516. On the basis of a license for exploitation, the holder is entitled to perform works necessary for establishing mines and quarries.[665] In addition, the license for exploitation establishes its duration; the exploitation perimeter, the quantity of reserves proposed to be exploited, the types of mining products, the holder's obligations according to the exploitation development plan and the environmental restoration plan, together with any other rights and obligations arising from the law.[666]

517. License holders are entitled to solicit the extension of the license's validity period, within the boundaries of the granted perimeter, by submitting to NAMR the required documentation,[667] including a report for the initial validity period of the license together with the programme for the supplementary works schedule for the additional period requested.[668] The request for extension must be submitted within minimum ninety days before the expiration of the initial license, otherwise the concession or the administration right will automatically expire,[669] as provided by the Mines' Law.

518. Where the holder of the license proves that the deposit stretches beyond the boundaries of the exploitation perimeter, then he is entitled to ask the extension of the perimeter, but only after the expiration date of the initial validity period of the license, or after the exhaustion of the coal resources/reserves from the initially granted perimeter. However, the extension will be denied where it overlaps with another concession or administration right already in place.[670]

519. The start of the mining activities established in the license is authorized in writing by the competent authority within 180 days since the entry into force of the license but no longer than 30 days after the holder has submitted cumulatively the following documents: the evidence of paying the mining activity tax, the evidence of posting the bond for environmental restoration, the certificate issued by the fiscal authorities, attesting that the holder does not have any outstanding obligations to the state, the environmental approval, the approval of the competent authority for the

---

664. Article 21 of Mines' Law.
665. Works are detailed in Art. 30 of the Guidelines for Mines' Law and include: construction and assembly of equipment and other necessary utilities, specific for extraction works, processing, transport, temporary storage of coal and waste products, surface and underground works for coal extraction, processing and delivery in specific forms as well as research works for increasing the level of knowledge concerning coal resources and reserves.
666. Article 31 of the Guidelines to Mines' Law.
667. Article 32, para. 1 of Guidelines to Mines' Law. Complete list of the required documentation is provided by Art. 20 of Mines' Law.
668. Article 32, para. 3 of the Guidelines to Mines' Law.
669. Article 31, letter a) of Mines' Law.
670. Article 32, para. 2 of the Guidelines to Mines' Law.

annual exploitation programme and the approval of the land's owner or administrator regarding the access to the surfaces needed for the performance of the mining activities contained in the annual exploitation programme.[671]

§2. Permit for Exploitation

520. The permit for exploitation is a legal deed issued by NAMR by which it grants the right of exploitation of a determined quantity of rocks usable in constructions,[672] accumulations of peat or gold from river deposits.[673] However, beside the limitation of the quantity, the law states that the right for exploitation based on a permit is granted for rocks, peat and gold which were either or not explored and registered as capitalized reserves or resources in the Mineral Resources National Fund and which do not constitute already the object of a concession or right of administration at the moment when the permit was requested.[674]

521. Permits are not granted for perimeters which are not on the list of perimeters for concession of exploration or exploitation activities but they may be granted to other persons than those holding a license for exploration or exploitation in place in a specific perimeter.[675]

522. Exploitation based on a permit may be performed by either a legal person, registered or organized according to the law, specialized and authorized for conducting such operations or to authorized and registered natural persons.[676]

523. The request for permit of exploitation can be submitted at any point throughout a calendar year and comprises of an order-letter for granting the permit and the technical documentation, on a first came first served basis.[677] The permit is valid for one year, without the right for extension and without granting the holder any right for direct concession or the right to obtain a new permit.[678] However, the permit shall remain valid for its entire period even if during its validity occur requests for concession of the exploration or exploitation activities and/or, as a result, NAMR decides to hold a public tender for offers concerning an area which includes the perimeter covered by the exploitation permit.[679]

---

671. Article 22 of Mines' Law.
672. The categories of rocks usable in constructions which are exploitable on a basis of a permit are: sand, gravel, marble, etc.
673. Article 1, para. 1 of NAMR's Guideline concerning issuance of permits for exploitation of 30 Apr. 2009 (hereinafter Guideline on mining permits), published in the Official Gazette of Romania, Part I, No. 333/19.05.2009, subsequently amended by Order No. 142/2010 and Order 125/2011.
674. Article 3, para. 1 of Guidelines on mining permits.
675. Article 3, paras 2 and 3 of Guidelines on mining permits.
676. Article 4 of Guidelines on mining permits.
677. Article 6 of Guidelines on mining permits.
678. Article 7 of Guidelines on mining permits.
679. Article 18 of Guidelines on mining permits.

524. The procedure for obtaining the permit is detailed in the Guidelines on mining permits and covers: the documents or authorizations required and the financial bond for environment restoration. Additionally, applicants for exploitation permits are obliged to pay towards the state budget the tax for the exploitation activity and the mining royalty according to the legal provision.[680]

525. Within thirty days from the moment it received the required documentation, NAMR, through its territorial inspection compartments, draws up the permit or declines the request. In case NAMR approves the request, the issued permit will mainly establish: the holder; the object; the validity period; the name and location of the perimeter; the type and the determined quantity of the mineral resource; the mining products; the rights and obligations of the holder; the reasons for termination; the taxes and royalties owed and their deadlines; the obligation to comply with: the Mines' Law, environmental protection rules, and the terms of approvals and authorizations concerning water management; the addendums to the permit; the right of NAMR to inspect whenever it deems necessary and to apply sanctions; and any other provisions and special covenants.[681]

526. The permit, together with the order-letter, the technical documentation and the proof of tax payment are submitted to NAMR for approval by its president. NAMR notifies the holders of the approval within three days from the registration of the permit, through registered mail with receipt confirmation.[682] The permit will be released to its holder only after the issuance tariff has been paid.[683]

§3. Cessation of Mining Operations: Closing the Mines

527. Exploitation activities of a deposit cease when: the mining resources are exhausted; the exploitation has become impossible due to natural causes (floods, collapses, landslides) or provoked events (endogenous fires or gas explosions) whose effects cannot be removed through technical operations which are economically viable; the exploitation became economically unprofitable. Whenever one of the aforementioned causes has occurred, the holder[684] must seek approval of temporary or permanent cessation from the competent authority (in case of perimeters granted to national mining companies the cessation must be approved first by the

680. Articles 12 and 13 of Guidelines on mining permits.
681. Article 16 of Guidelines on mining permits.
682. Article 17, paras 1–2 and 4 of Guidelines on mining permits.
683. According to Art. 17, para. 3 of Guidelines on mining permits the tariff must be paid by the holder in sixty days from the notification of approval, under the sanction of cancellation.
684. Article 52, para. 1 of Mines' Law states that the initiative for cessation of exploitation belongs to the license holder who must submit a request accompanied by an updated plan of cessation, containing: the reasons for cessation, the technical programme for decommissioning and maintenance of the operation, including a plan for post-closure monitoring; a programme for social protection of personnel, drawn up according to the law, after consultation with the affected constituencies and approved by the competent authority in the area of social protection; an authorization for water management and an environmental authorization for closing the mine or the pit in question; and the procedure for decommissioning and release of the land.

relevant minister).[685] The decision to close or maintain the mines belonging to the national mining companies must be approved by the government given that the expenses generated by the closing or maintenance will be co-financed from budgetary sources.[686]

528. According to the law, the mining perimeters with mining activity which is temporarily ceased may be given into concession to a different holder, as a result of a new public tender for offers.[687]

685. Article 51 and Art. 52, para. 2 of Mines' Law. The cessation takes place by decision of the competent authority (and of the relevant minister, as the case may be), after the analysis and acceptance of the closing plan and its implementation. The supervision of compliance with the obligations resulting from the closing of the mine lies with the competent authority or the relevant ministry, as the case may be, as stated in Art. 53, para. 1 of Mines' Law.
686. Article 52, para. 2, Second thesis of Mines' Law.
687. Article 52, para. 4 of Mines' Law.

Chapter 6. Distribution and Consumption

529. According to available data Romanian coal reserves are nearly exhausted. The known reserves' figure is 291 million tons, which would ensure the country's necessities for only nine years, if the current production conditions are maintained.[688] However, due to policies concerning the reduction of emissions, coal consumption in EU countries is reduced annually by 2.5% until 2035, therefore in the long run, coal reserves exhaustion might not have such a significant impact on the Romanian energy sector. For instance, while in 2006, Romanian coal production was 6.5 million tonnes oil equivalent, in 2015 it dropped to 4.7 million tonnes oil equivalent, respectively, 4.3 million tonnes oil equivalent in 2016. Thus, there was a decrease of 9.2% in the total production of coal at national level between 2015 and 2016.[689]

688. Harta Resurselor Romaniei. In cati ani se vor termina rezervele de gaze natural, petrol si carbune, 30 Dec. 2013, gandul.info, available online, in Romanian only, at: http://www.gandul.info/financiar/harta-resurselor-romaniei-in-cati-ani-se-vor-termina-rezervele-de-gaze-naturale-petrol-si-carbune-11748644, last visited 23 Mar. 2018. The article was based on statistic data provided by BP Statistical Review of World Energy. The latest version of the report, from June 2017 can be found here: https://www.bp.com/content/dam/bp/en/corporate/pdf/energy-economics/statistical-review-2017/bp-statistical-review-of-world-energy-2017-full-report.pdf, last visited 23 Mar. 2018.

689. *Ibid.*, p. 38.

Part V. Interaction Between Energy Law and Environmental Law

530. According to a UN Report from 2001, reviewing Romania's Environmental Performance,[690] the energy sector was 'by far the largest contributor to air pollution in Romania'.[691] The main causes identified for this situation were low energy efficiency in all sectors due to 'the use of inefficient equipment and general lack of investment in modern technology'. However, the potential for energy efficiency improvement was deemed 'very high',[692] and recent monitoring data coming from the European Environmental Agency indicate a relatively steady decline in air pollution.[693]

690. Economic Commission for Europe, Committee on Environmental Policy – Environmental Performance Reviews: Romania, published by the United Nations, 2001, available online at: http://www.unece.org/fileadmin/DAM/env/epr/epr_studies/romania.pdf, last visited on 23 Mar. 2015.
691. *Ibid.*, at 187.
692. *Ibid.*, at 188.
693. https://www.eea.europa.eu/themes/air/country-fact-sheets/romania, last visited on 23 Mar. 2018.

Chapter 1. General Introduction

531. From a political standpoint Romania has aligned the EU policies concerning the environment, through the obligations assumed in the prior- and post-accession period. The Accession Treaty signed by Romania with the EU in 2005 provided that Romania must implement the entire *acquis*, with several delayed deadlines. In order to comply with its obligations Romania adopted entire packages of policies and legislation concerning environmental protection, covering air, water and soil quality, pollution, risk management, waste management and disposal or climate change. The general framework is represented by Government Emergency Ordinance (GEO) No. 195/2005 on environmental protection,[694] which implemented Directive 96/61/EC on the integrated prevention and control.

§1. Pollution Prevention

532. GEO No. 195/2005 sets out the necessary measures regarding prevention or, at least reduction, of air, water and soil pollution resulting from activities with a significant impact on the environment (such as production of energy), including measures concerning waste management. Therefore, in order to carry out the aforementioned activities, an integrated environmental authorization is required. It is solely based on such authorization that one can be granted a right to operate an energy- or mining-related installation.

533. The environmental authorization contains the measures, requirements and conditions that must be complied with in order to carry out activities with a significant environmental impact. In order to ensure control over the operator's compliance the law does not allow any substantial change in the operation of an installation without the prior approval of the competent authority. Where such situation arises, the environmental authority reviews the terms and conditions of the authorization and decides whether there is a need for adjustments and issues amended requirements or conditions for operating the installation under the new circumstances. However, where a substantial change in the operation of the installation has occurred, the operator is obliged to carry out an environmental impact assessment and obtain the appropriate approvals required by such change.

§2. General Principles of Environmental Protection

534. Romanian environmental protection is based on a series of principles and strategic elements such as:

– The principle of integrated environmental conditions in all sector policies.

694. Government Emergency Ordinance No. 195/2005 concerning environmental protection (hereinafter GEO 195/2005) was published in the Official Gazette of Romania, Part I, No. 1196/30.12.2005 and was subsequently amended in 2006, 2007, 2008, 2011, 2012 and 2016.

- The principle of precaution in decision making.
- The principle of preventive action.
- The principle of retaining polluters at the source.
- The principle of 'polluter pays'.
- The principle of preserving bio-diversity and ecosystems specific to the natural bio-geographical frame.
- The principle of sustainable use of natural resources.
- The principle of informing of and involving the public in decision making, as well as access to justice in matters concerning environmental issues.
- The principle of developing international cooperation concerning environmental protection.[695]

535. In order to ensure implantation of the aforementioned principles, Romanian law states a series of methods: prevention and integrated pollution control by employing the best available techniques for activities with a significant impact on the environment; correlation of spatial and urban planning with environmental planning; conducting environmental evaluation before approving plans and programmes which might have a significant impact on the environment; introducing and utilizing economic instruments and leverages for stimulation or coercion; establishment of a national system for integrated monitoring of environmental quality; acknowledging products with a reduced impact over the environment by granting ecological labelling; or restoration of areas affected by pollution.[696]

§3. Liability for Activities which Affect the Environment. Obligations of Natural and Legal Persons

536. By law, environmental protection is the obligation and responsibility of both local or central public administration and all natural or legal persons.[697] In order to comply with these obligations,[698] natural and legal persons carrying out activities in Romania must apply for and abide by environmental permits or authorizations to the environmental protection authorities;[699] must pay for damages caused by pollution and restore the environment;[700] must prevent and avoid technological risks or accidental pollution by monitoring, analysing and recording their technological installations and processes and report the results to the competent authorities;[701] collaborate with the authorized personnel for control and comply with the measures imposed by the competent authorities.[702]

695. Article 3 of GEO No. 195/2005.
696. For a complete list of methods of implementing of principles and strategic elements of environmental protection *see* Art. 4 of GEO No. 195/2005.
697. Article 6, para. 1 of GEO No. 195/2005.
698. A full list of obligations of natural and legal persons is to be found in Art. 9455 of GEO No. 195/2005.
699. Article 94, para. 1, letter a) and b) of GEO No. 195/2005.
700. Article 94, para. 1, letter i) of GEO No. 195/2005.
701. Article 94, para. 1, letter j)-l) of GEO No. 195/2005.
702. Article 94, para. 1, letter e)-h) of GEO No. 195/2005.

I. Specific Obligations for Natural and Legal Persons Conducting Energy-Related Operations

537. Specific obligations for natural or legal persons which conduct prospections, explorations or exploit the natural resources of soil and subsoil refer to:

(1) Restoration of areas where the soil, subsoil and terrestrial ecosystems have been affected.
(2) Notification of environmental protection authorities or any other competent administrative bodies, as the case may be, on any accidents which might endanger the environment and to act in order to restore it.[703]

II. General Regime of Liability: 'Polluter Must Pay' Principle

538. Notwithstanding whether the activity causing a significant impact on the environment is related to energy or not, the liability for damages cause to the environment is the same since it has an objective character[704] (meaning that it is independent of polluter's fault). In addition, the law specifies that in case of plurality of authors, the liability will be joint.[705]

539. According to the law, the prevention and compensation for damages caused to the environment shall take place in accordance with the provisions of GEO 195/2005 and any other specific regulations. As shown earlier, Article 3, letter e) has established the 'polluter must pay' principle, therefore, those legal or natural persons that cause any environmental damage are bound to pay all costs pertaining to the damage remedy, the removal of its consequences as well as any costs needed to restore the previous conditions.

540. The details are set out in GEO No. 68/2007[706] which implemented in Romania the provisions of the EU Directive No. 2004/35/CE concerning environmental liability with focus on prevention and remedy of environmental damages. However, GEO 68/2007 specifically states that it does not provide private law legal and natural persons with a right to remedy as consequence of a damage to the environment or an imminent danger caused by environmental damage. In such instances, the liability shall be established in accordance to the provisions of general civil law.[707]

703. Article 94, para. 2 of GEO No. 195/2005.
704. For exceptions from the objective character of liability *see* Art. 95, para. 2 of GEO No. 195/2005.
705. Article 95, para. 1 of GEO No. 195/2005.
706. Government Emergency Ordinance No. 68/2007 (hereinafter GEO No. 68/2007) was published in the Official Gazette of Romania, Part I, No. 446/29.06.2007 and was subsequently amended in 2009, 2011, 2012 and 2013.
707. Article 3, para. 4 corroborated with Art. 44 of GEO No. 68/2007.

541. The authority in charge with the assessment of the environmental damage, the necessary reparatory measures and the evaluation of damages is the county environmental protection agency. The same institution is required to consult the local county offices of the Environment National Guard (ENG) or other authorities and public institutions.[708] However, the competent authority for establishing the damage of the environment or the imminent threat to cause such damage, as well as identifying the responsible operator is the ENG through its county offices.[709]

542. The decisions of the county agency's president concerning the establishment of pre-emptive or reparatory measures are motivated and contain information concerning the deadlines and the procedure for administrative litigation as of Law No. 554/2004.[710]

543. GEO No. 68/2007 further sets out the pre-emptive[711] and the reparatory[712] measures to be taken and provides a right to action to any natural or legal person which is or could be affected by a damage to the environment. Therefore, such person is entitled to petition the relevant authority, in writing or through electronic means, with a request to take all legal measures or to transmit to the competent authority any observations concerning the occurrence or imminence of a damage to the environment. However, as already stated, this right to action does not entitle legal or natural persons to claim compensations. The petitioned agency must decide whether to act or not upon the request received and inform the petitioner thereof. The decision may be challenged in administrative courts.[713]

544. GEO No. 68/2007 restates and details the 'polluter pays' principle[714] and sets the exceptions from its application.[715] As established in the general framework concerning environmental protection, in case the damage is caused by more than one operator, the liability is joint.[716] The right of the county authority for environmental protection to seek damages shall be prescribed in five years since the moment the reparatory measures have been taken or since the date when the responsible operator has been identified.[717]

545. Despite the general principle of polluter pays, the owner of an asset or location may also be held liable in case of previous contaminations related to the said asset or location. As it is pointed out,[718] the rules are not clear on this matter for

708. Article 6 of GEO No. 68/2007.
709. Article 9 of GEO No. 68/2007.
710. Article 8 of GEO No. 68/2007.
711. Articles10–12 of GEO No. 68/2007.
712. Articles 13–19 of GEO No. 68/2007.
713. Articles 20, 21, 23 and 25 of GEO No. 68/2007.
714. Article 26 of GEO No. 68/2007.
715. Article 28 of GEO No. 68/2007.
716. Article 31 of GEO No. 68/2007.
717. Article 32 of GEO No. 68/2007.
718. Roxana Ionescu, *An Overview on Environmental Law in Romania, PLC Cross-border Environment Handbook 2008–2009*, available online at http://www.nndkp.ro/publications/articles/environment-handbook_/, last visited on 23 Mar. 2018.

which reason one must look for additional support in the general framework established by GEO 195/2005, from whose provisions it can be inferred that in cases where the polluter cannot be identified, the liability lies with the owner of the location. Hence it was argued that the pollution of a location is in fact a risk of the asset.[719] Under the general regime of the NCC, this would mean that the transfer of the property over that asset would entail the transfer of all associated risks, unless the contracting parties have specifically excluded them or if the risks were hidden.

719. *Ibid.*

Chapter 2. Authorities with Competences in Environmental Protection: Powers and Duties

546. Currently, upon enforcing the EU *acquis* in the field of environmental protection, there are several competent authorities responsible for environmental protection. According to Article 2, point 12, these are: (a) the central public authority for environment protection – the Ministry of Environment, Waters and Forests; (b) the National Agency for Environmental Protection (NAEP); (c) the county environmental protection agencies; (d) the Administration of Danube Delta Biosphere (ADDB) and (e) the Environment National Guard (ENG) with its subordinated structures. Since environmental protection is, as already stated, an obligation of all central and public authorities, the environmental authorities are supported in their activities by all other public, central and local administrative bodies.[720]

547. The control of compliance with the measures concerning environmental protection is realized by the inspectors and authorized personnel of ENG and ADDB, the authorized personnel of local public bodies and by the authorized personnel of the National Commission for Controlling Nuclear Activities; the Ministry of National Defense and the Ministry of Internal Affairs (but only within their area of activity).[721]

§1. Powers and Duties of the Central Public Authority

548. The Ministry of Environment, Water and Forests enjoys a large number of powers, doubled by a large set of duties granted by GEO 195/2005. As part of its set of duties, the central authority: updates the environmental protection strategy; organizes the national integrated monitoring system of environmental quality, coordinates its activity and informs the central authority for health with respect to the results of monitoring radioactive contamination of the environment; implements the policies, strategies and regulations for environmental protection through NAEP and county authorities; collaborates with local public bodies and other legal persons for the purpose of restraining the negative effects of economic activities on the environment and advocates for implementing techniques and technologies appropriate for the environment; collaborates with local public bodies of civil protection in the elaboration of operative plans and common performance of interventions in case of contamination or ecologic accidents.[722]

549. Among the powers granted by the law, the Ministry enjoys the right: to initiate normative acts, guidelines, regulations and specific procedures; to approve guidelines and regulations concerning activities with impact on the environment which were elaborated by other authorities and to control their implementation; to control, within its competence sphere the compliance with obligations assumed

720. Article 6 of GEO No. 195/2005.
721. Article 7, para. 3 of GEO No. 195/2005.
722. For a full list of duties *see* Art. 75 of GEO No. 195/2005.

through international conventions or those established by national law; impose sanctions through ENG for breaches of environmental legislation or non-compliance with the conditions established through regulations; and to inspect and control through the ENG any activities related to environmental protection.[723]

550. NAEP is a public body having its own legal personality, funded entirely from the state's budget. It is the specialized authority for implementing the policies and legislation concerning environmental protection, but it is entirely subordinated to the central public authority. Similarly, all competent authorities for environmental protection are public bodies with legal personality, financed by the state's budget.

551. Among the legal powers granted to these authorities, the law specified the right of all competent authorities to request necessary data from other central or local public bodies, or from natural and legal persons.

552. With respect to the investigations performed by the chief inspectors and the inspectors of ENG and other authorized personnel have access, within the confinements of the law, at any time and in any location where an activity having an impact over the environment is carried out.[724]

§2. Powers and Duties of Other Central and Local Authorities

553. Central public bodies bear the following obligations: to elaborate the policies and strategies concerning their own area of activity or sectorial plans and programmes in compliance with the provisions of environmental protection legislation; to integrate environmental policies in their specific policies; or to prepare and communicate to the competent authority for environmental protection all data and documentation required for obtaining the environmental authorization for the plans and programmes for which an environmental assessment is required by law.[725]

554. Further specific duties and obligations are set out for the public authorities with competences in the area of development and forecast,[726] the central public authority for health,[727] the National Ministry for Defense or the central public bodies acting in the area of national security[728] and internal affairs,[729] the central public

723. For a full list of duties *see* Art. 75 of GEO nor 195/2005.
724. Article 79, para. 1 of GEO No. 195/2005. Exceptions from the general rule of unlimited access are specified in paras 4 and 5.
725. For a full list of obligations *see* Art. 80 of GEO No. 195/2005.
726. Article 81 of GEO No. 195/2005.
727. Article 82 of GEO No. 195/2005.
728. Articles 83–84 of GEO No. 195/2005.
729. Article 89 of GEO No. 195/2005.

bodies for education and research,[730] economy and trade,[731] as well as those for agriculture, forest and rural development[732] or for transport, construction and tourism.[733]

555. The Police, the county police, the Inspectorates for Emergency Situations and the Financial Guard are obliged to provide support, by request, to the representatives of environmental protection competent authorities during the performance of their duties.[734]

§3. Taxes and Tariffs

556. Public environmental protection authorities are entitled to perceive tax for issuing the authorizations or permits established by the law. These taxes are part of the Environmental Fund and their amount is set by Government Decision, at the proposal of the central public authority.[735]

557. Additionally, the competent environment authorities may perceive tariffs for works and services performed at the request of legal and natural persons, as part of their legal duties. The amount of tariffs is set by Order of the central public authority's head. Amounts obtained from tariffs constitute revenues to the state's budget.[736]

730. Article 85 of GEO No. 195/2005.
731. Article 86 of GEO No. 195/2005.
732. Article 87 of GEO No. 195/2005.
733. Article 88 of GEO No. 195/2005.
734. Article 93 of GEO No. 195/2005.
735. Article 22 of GEO No. 195/2005.
736. Article 23 of GEO No. 195/2005.

Chapter 3. Authorization of Activities Having an Impact on the Environment

558. The law stated an obligation to apply for the issuance of an environmental permit or authorization for any new activities that have a potential significant impact on the environment,[737] or for amending or extending existing activities that might have the same effects.[738] Performing existing activities or starting new ones may be performed solely on the basis of an environmental authorization or environmental integrated authorization.[739] The social and economic activities requiring environmental approval are expressly listed by law.[740] The procedure for authorization is public.

§1. Environmental Permit and Authorization

559. The environmental permit is defined as the administrative deed issued by the competent environmental authority which establishes the conditions or the measures (as the case may be) for environmental protection that need to be observed for the implementation of a project.[741]

560. The environmental authorization is defined as the technical and legal document which establishes the operating conditions and parameters for either an existent or a new activity with a potential significant impact on the environment. The legal definition adds that having an authorization at the moment of start-up of activities is mandatory.[742] Therefore, according to Romanian legislation, activities can be conducted only after the issuance of the environmental authorization, any violation being sanctioned with an administrative fine.[743]

561. All administrative documents may be revised if new elements or elements which were unknown at the time of issuance have occurred or when conditions which were in place at the time of issuance have changed. Therefore, in situations

737. Article 9, para. 1 of GEO No. 195/2005.
738. Article 11 of GEO No. 195/2005.
739. Article 12, para. 1 of GEO No. 195/2005. Whether an operator requires an integrated authorization or just an authorization depends on the type of activity performed. Integrated environmental authorizations fell under the scope of Government Emergency Ordinance No. 152/2005, published in the Official Gazette of Romania, Par I, No. 1078/30.11.2005 which has transposed into national legislation the EU Directive 2008/1/EC concerning integrated pollution prevention and control. GEO No. 152/2005 was repealed and replaced by Law No. 278/2013 concerning industrial emissions, published in the Official Gazette of Romania, Part I, No. 671/01.11.2013 (hereinafter Law No. 278/2013) and amended in 2017, by GEO No. 101/2017, published in the Official Gazette of Romania, Part I, No. 1007/2017, for the modification and completion of Law No. 278/2013 (hereinafter GEO No. 101/2017).
740. Article 12, paras 2 and 13, para. 1 of GEO No. 195/2005.
741. Article 2, point 3 of GEO No. 195/2005.
742. Article 2, point 9 of GEO No. 195/2005.
743. Article 14 corroborated with Art. 96, para. 3, point 1 of GEO No. 195/2005.

where operators of activities which require the position of the environmental protection authorities are undergoing a procedure for sale of the control stock, the sale of assets, a merger, division, concession or other situations involving changing the ownership of the business, and in case of dissolution followed by liquidation, insolvency, cessation of activity, the operators must notify the competent authority thereof and obtain its approval before the change.[744]

562. The validity of the environmental permit lasts throughout the carrying out of the plan, programme or project. However, the environmental authorization is valid for five years, while the integrated one is valid for ten.[745]

§2. Environmental Impact Assessment: Stages – Types of Assessment

563. In order to obtain or amend the environmental permit or authorization, the operators must submit to the competent authority, together with other documents, an environmental impact assessment. These assessments do not have the legal value of a permit or license, but are mere prerequisites in the procedure for the issuance of the permit or authorization. The environmental impact assessment process is part of the procedure established by Government Decision No. 445/2009 on environmental impact assessment regarding specific public and private projects, which transposed into national law the provisions of EU Directive 85/337/EEC on the assessment of the effects of certain public and private projects on the environment,[746] and its Guidelines approved by Order No. 135/2010 of the Minister for Environment, Waters and Forests.[747]

564. The environmental impact assessment process is preceded by a preliminary stage at which the county environmental authority establishes the geographical setting of the project in relation to the protected natural areas of public interest.

565. The preliminary stage starts with the operator's application to obtain an environmental approval for carrying out the project. The county authority establishes if the project needs to be subjected to an environmental impact assessment. Where the case may be, the county authority informs the operator of its decision and requests that the operator drafts a document containing the project activities that have to be assessed. The document is further submitted by the county authority to the competent environmental authority for carrying out the assessment.[748]

---

744. Article 15, paras 1 and 2 corroborated with Art. 10, para. 1 of GEO No. 195/2005.
745. Article 16, paras 1 and 2 of GEO No. 195/2005.
746. Government Decision (GD) No. 445/2009 on environmental impact assessment regarding specific public and private projects was published in the Official Gazette of Romania, Part I, No. 481/2009, subsequently amended in 2012.
747. Order No. 135/2010 of the Minister for Environment, Water and Forests, concerning the approval of the Guidelines for assessment of environmental impact for public and private projects, was published in the Official Gazette of Romania, Part I, No. 274/27.04.2010.
748. Articles 8–10 of Order No. 135/2010.

566. After the preliminary stage is completed, the environmental assessment procedure goes on in three main phases. In the first phase, the competent authority analyses the aforementioned document, establishes a technical analysis committee, assesses whether the project has a cross-border dimension and informs interested third parties about the project owner's application. This phase is finalized with the competent authority's decision either to perform or not an environmental impact assessment. The applicant is informed thereof, and interested third parties are also informed through mass communication channels. Third parties may express their points of view and amendment suggestions with regard to the decision of the authority, which may lead to a re-evaluation of the decision by the issuing authority.[749]

567. In the second phase, the competent authority identifies the points to be addressed under the environmental assessment and details them in a guide aimed at helping the operator to draft an environmental impact assessment report. The report has to be drafted by an authorized expert. The points to be addressed in the environmental assessment report represent the potential effects on the environment generated by the project and the solutions thereto. The report has to consider suggestions of the interested parties.[750]

568. In the third phase, the report is released for public debate and the quality of the report is analysed. It is the obligation of the applicant to organize the public debate and cover its costs. After the debate, the operator has to prepare answers and solutions to the issues raised by the participating third parties (which will be annexed to the report) and also has to perform any amendments to the report as requested by the technical analysis committee.

569. After analysing the amendments (where the case may be), the competent authority either issues the environmental approval or rejects the application. The operator then informs any third parties of the environment authority's decision by mass communication channels.[751]

570. Under Romanian legislation, members of the interested public (i.e., citizens, NGOs or any structure or entity that may be affected by the implementation of a project) have the right to be actively involved in the environmental impact assessment process from the submission of the request for issuance of the environmental agreement until the relevant authority's final decision.[752]

571. On average, the environmental impact assessment process takes between three and six months, but the procedure can be longer.

749. Article 11–12 of Order No. 135/2010.
750. Article 13–16 of Order No. 135/2010.
751. Article 17–25 of Order No. 135/2010.
752. Article 16 of GD No. 445/2009 concerning the evaluation of the environmental impact of certain public and private projects, published in the Official Gazette of Romania, Part I, No. 481/13.07.2009, later amended by GD No. 17/2012 for the modification and completion of GD No. 445/2009, published in the Official Gazette of Romania, Part I, No. 48/20.01.2012 (hereinafter GD No. 445/2009).

572. Environmental assessments may be: environment assessment level 0; environment assessment level I; and environment assessment level II. The type of assessment necessary to grant a permit/authorization for a project/activity is either stipulated by the regulations in force, or it may be required by the competent environmental authority, entitled to issue the required permit/authorization, depending on the environmental impact of the location. The three types of environment assessment do not exclude each other and can be drawn up consecutively or simultaneously.

573. A level 0 environmental assessment is drawn up whenever such assessment is required by law, without indicating its type, and the environment authority considers less probable the existence of an environmental impact of the location. The level 0 assessment represents the minimum requirements for situations where, according to the law, presentation of an environmental assessment is necessary.

574. The level I environmental assessment will be required by the environment authorities when: level 0 assessment shows a possible environmental impact, on that location; the current activity of an entity fails to meet the authorization conditions; or upon changing the investment destination or owner, or in case of cessation of the activities generating an environment impact. Environment assessment level II will be required by the environment authority when: the level I assessment indicates a potentially significant pollution of a location; or the environmental authority establishes from the inception of level I assessment that the location presents a potentially significant pollution.

575. The level II environmental assessment is the only assessment requiring sample taking and physical, chemical or biological tests of the environment factors (air, water, soil, etc.) The level I and II environmental assessment and impact studies can only be elaborated by specialized entities or individuals, certified according to the laws in force.

Chapter 4. Areas of Protection

§1. Regulation Concerning Waste

576. The general framework concerning waste is contained in GEO 195/2005, Articles 29–33, which states that the control of waste management belongs to competent public bodies for environmental protection and other authorities with competences established by the legislation into force.

577. In 2011, Romania adopted Law No. 211/2011[753] which transposed into national legislation the EU Directive 2008/98/EC on waste. Unlike the previous normative act, Law No. 211/2011 does not contain a list of waste categories but provides a general definition. The list of waste is defined by the European Commission, although objects or substances included on that list are not automatically considered to be waste. However, the law is not applicable to radioactive waste or to waste which is covered by other normative acts. Therefore, waste resulting from prospection, extraction, treatment and storage of mineral resources or the working of quarries fall outside its scope.[754]

§2. Regulation of Air Emissions: Protection of Atmosphere and Climate Change

578. The general regime is again set in GEO No. 195/2005, Articles 59–64 which states the main duties and powers of the central public authority for environment protection among which those relevant to the energy sector refer to providing the limits for air emissions and concentration in the air[755] and proposing the Scheme for Trading of Greenhouse Gas Emissions Certificates (GGEC).[756]

579. The Scheme has relevance to the field of energy law for it applies to emissions deriving from activities such as: combustion installations with a rated thermal input exceeding 20 MW (except for hazardous or municipal waste installations); mineral oil refineries; and coke ovens.

753. Law No. 211/2011 concerning the waste regime was republished in the Official Gazette of Romania, Part I, No. 220/28.03.2014 (hereinafter Law No. 211/2011), was amended by GEO No. 68/2016 for the modification and completion of Law No. 211/2011, published in the Official Gazette of Romania, Part I, No. 823/18.10.2016 (hereiafter GEO No. 68/2016). In its turn, GEO No. 68/2016, was amended by Law No. 166/2017, concerning the approval of GEO No. 68/2016 for the modification and completion of Law No. 211/2011 concerning the waste regime, published in the Official Gazette of Romani, Part I, No. 554/13.07.2017.

754. Article 1, para. 2, letter d) of Law No. 211/2011.

755. Article 59, letter l) of GEO No. 195/2005.

756. Article 60, para. 1 of GEO No. 195/2005.

580. In 2011, Romania adopted Law No. 104/2011 on ambient air quality[757] which transposed into national legislation both EU Directive 2008/50/EC on ambient air quality and cleaner air for Europe and EU Directive 2004/107/EC related to arsenic, cadmium, mercury, nickel and polycyclic aromatic hydrocarbons in ambient air.

§3. Protection of Soil, Subsoil and Terrestrial Ecosystems

581. By law the protection of soil, subsoil and terrestrial ecosystems is mandatory to all land holders, notwithstanding their title.[758] Furthermore, land holders under any title, as well as any natural or legal person who carry out activities on a certain piece of land without having a legal title are bound by the following obligations: to prevent damage to the quality of geological environment; to ensure the cleaning up on the lands which are not subject to production or operation, especially those located along roads, railways or shipping communication routes; and to observe any other obligations arising from applicable additional regulations.[759]

§4. Protection of Fresh Water and Seawater

582. The legal framework concerning water management is established by Law No. 107/1996.[760] According to its provisions, the right to use water is granted through the authorization for management of water which also includes the right to discharge wastewater. Individuals are free to use for domestic purposes seawater and internal water, outside restricted areas.

583. By Order 662/2006 of Ministry of Environment, Water and Forests, governing the procedure and authority to issue water management permits and authorizations, it is established that the responsibility for granting such permits to operators carrying out water-related activities, lies with the national administration and its subordinated units.[761]

757. Law No. 104/2011 on ambient air quality was published in the Official Gazette of Romania, Part I, No. 452/2011 (hereinafter Law No. 104/2011), and was amended by GD No. 336/2015 concerning the amendment of Annexes 4 and 5 to Law No. 104/2011, published in the Official Gazette of Romania, Part I, No. 343/19.05.2015, and by GD No. 806/2016 for the amendment of Annexes 4, 5, 6 and 7 to Law No. 104/2011, published in the Official Gazette of Romnaia, Part I No. 898/09.11.2016.
758. Article 65 of GEO No. 195/2005.
759. Article 68 of GEO No. 195/2005.
760. Law No. 107/1996 on waters, was published in the Official Gazette of Romania, Part I, No. 244/08.10.1996 and was subsequently amended in 1997, 1999, 2001, 2002, 2004, 2006, 2007, 2010, 2011, 2012, 2013, 2014, 2015, 2016 and 2017.
761. Order No. 662/2006 concerning the approval of the procedure and authority to issue water management permits and authorizations was published in the Official Gazette of Romania, Part I, No. 661/01.08.2006.

584. The competent authority for issuing permits and authorizations establishes the limits for discharge and the integrated monitoring of the discharge of waste water. Extraction activities thus require a water management permit or authorization, if the water flow for new operations exceeds two litres per second and water flow for provisional installations exceeds ten litres per second, and the resulting waste affects the water's quality.

§5. Sanctions and Remedies

585. Failure to comply with the legal obligations imposed by the law for operators of activities having an impact on the environment represents a misdemeanour and is sanctioned with a fine. The amount of the fine depends on the seriousness of the breach and damage that occurred.[762] Besides fines, violations of the legal provisions may also attract suspension[763] or (in cases where the reasons for suspension have not been remedied) cancellation of environment authorizations or permits, after a sixty days notification period,[764] as well as the cessation of activities. Last but not least, the law also provides for criminal liability.[765]

762. Article 96 of GEO No. 195/2005.
763. Article 17, para. 3 of GEO nor 195/2005.
764. Article 17, para. 4 of GEO No. 195/2005.
765. Articles 98–99 of GEO No. 195/2005.

Part VI. Interaction Between Energy Law and Tax Law

586. Energy taxation is a major source of financing for any country. If in 2014, Romania obtained RON 1.5 billion (about EUR 340 million) from revenues on oil and gas and mineral resources, excise taxes for energy products have generated about RON 12.3 billion (about EUR 2.8 billion). However, Romania maintains one of the lowest levels of environmental taxes (excise tax) in the EU.[766]

766. Centrul de Politici Durabile – Analiza taxelor de mediu in Romania, pp. 18–19, available online in Romanian only at: http://www.ecopolis.org.ro/files/Analiza%20taxelor%20de%20mediu%20in%20Romania.pdf, last visited 23 Mar. 2018.

Chapter 1. General Introduction

587. The current Romanian taxation regime, including energy taxation, is provided mainly by the Fiscal Code,[767] together with all Guidelines and Instructions for its application[768] and other specific legislation, such as Government Ordinance No. 5/2013 concerning the establishment of special taxation measures for natural monopoly activities related to electricity and natural gas sector.[769] Since Romania became an EU member in 2007, its legislation is now harmonized with the EU *acquis* with regard to excise taxation of mineral oils, electricity, natural gas and coal.

588. With respect to energy taxation, Romania provides for three main types of taxes: income tax, VAT and excise tax,[770] which will be each later analysed solely with respect to energy and energy products. The Fiscal Code also provides for taxes levied by local authorities and taxes for constructions.

589. Taxation and fiscal measures are based on the following legal principles: *neutrality* with respect to the different categories of investors and capital, property type, thus ensuring equal conditions to investors and foreign or national capital; *certainty of taxation*, by elaborating clear legal norms, which cannot lead to arbitrary interpretations, and establishing clear deadlines, procedures and amounts to be paid for each taxpayer, so that they are able to follow and understand their allotted tax burden; *fairness* or *tax equity* for each taxpayer, based on the size of their income or assets; *efficiency of taxation* to ensure similar levels of budgetary income, by maintaining the yield of taxes, fees and contributions in all stages of the economic cycle, both during periods of economic growth and crisis; and *predictability of taxation* to ensure stable mandatory taxes, fees and contributions, for a period of at least one year, during which no changes can occur, in the sense of increasing or introducing new levies.[771]

767. The Fiscal Code was published in the Official Gazette of Romania, Part I, No. 688/10.09.2015 and was subsequently amended every year, including 2018.
768. According to Art. 5 of the Fiscal Code, the responsibility of unitary application of its provisions lies with the Ministry of Public Finances.
769. Government Ordinance No. 5/2013 concerning the establishment of special taxation measures for natural monopoly activities related to electricity and natural gas sector (hereinafter GO No. 5/2013) was published in the Official Gazette of Romania, Part I, No. 52/23.01.2013, was subsequently amended in 2014, 2015, 2016, 2017, and it will be into force only until 31 Dec. 2018.
770. Article 2, para. 1 of the Fiscal Code.
771. Article 3 of the Fiscal Code.

Chapter 2. Tax on Income or Profits Obtained in Romania

590. According to the fiscal legislation, the following categories of income are deemed to be obtained in Romania, notwithstanding whether they are received in Romania or abroad: income/profits attributable to a permanent establishment in Romania, income/profits from exploitation of natural resources situated in Romania, or any other activity performed in Romania, excluding transport services or services ancillary to transport.[772]

591. The income and profit taxes mentioned above are owed by: Romanian legal persons; foreign legal persons who undertake activities through one or more permanent establishments in Romania; or non-resident legal and natural foreign persons who undertake activities in Romania, through an association, with or without legal personality.[773] The tax is levied as follows: in the case of Romanian legal persons or legal persons with registered office in Romania, established in accordance with European legislation, on their taxable profits, obtained from any source, both in and outside Romania; in the case of foreign legal persons who undertake activities through one or more permanent establishments in Romania, over the taxable profits made by the permanent establishment or, respectively, over the taxable office designated to fulfil the fiscal obligations; and, in the case of foreign legal persons whose income is generated by exploitation of natural resources obtained in Romania, over the corresponding part of taxable.[774]

592. The Romanian income/profit tax rate is 16%.[775] The taxable profit is calculated as the difference between the incomes earned from any source and all expenses incurred for the purpose of making profits, from which one deducts the non-taxable incomes[776] and to which it adds the non-deductible expenses.[777] For the calculation of the taxable income, the law provides tax incentives for research and development activities: a supplementary deduction of 50% of eligible expenses for these activities and the application of the accelerated amortization method for equipment and appliances destined to these activities.[778] However, research and

772. Article 12, letters a), fh) and k) of the Fiscal Code.
773. Article 13 of the Fiscal Code.
774. Article 14 of the Fiscal Code.
775. Article 17 of the Fiscal Code.
776. Among deductible expenses for the purpose of making profits, the Fiscal Code mentions those incurred for the protection of the environment and preservation of resources. *See* Art. 26, para. 2 and 3 of the Fiscal Code.
777. Article 25, para. 4, of the Fiscal Code corroborated with Points 16–17 of the Guidelines for the Application of Law No. 227/2015 concerning the Fiscal Code of 6. Jan. 2016 (hereinafter Guideline to the Fiscal Code), published in the Official Gazette of Romania, Part I, No. 22/13.01.2016, with the subsequent amendments (the latest one from 2017).
778. Article 20, para. 1 of the Fiscal Code corroborated with the Guidelines concerning deductions for research and development expenses during determination of taxable profits of 5 Jul. 2016, published in the Official Gazette of Romania, Part I, No. 526/13.07.2016.

development activities do not include prospection, exploration, or production of fuel, oil or natural gas for the purpose of the Fiscal Code.[779]

593. Of interest to undertakings in the energy sector are also the provisions related to fiscal amortization. Thus, according to Romanian law, expenses related to the acquisition, production, construction, assembly, installation, or improving fixed assets are subjected to amortization. For an asset to be amortizable, the law established three cumulative conditions: the asset is possessed and used in the production or delivery of goods or in the performance of services; it has a fiscal value bigger than the limit established by Government Decision at the time when it becomes part of the taxpayer's patrimony; its usage period is higher than one year. Additionally, the law deems among amortizable fixed assets all investments performed for stripping related to the capitalization of mineral resources, as well as for opening works and works for preparing underground and surface extraction.[780]

594. The same regime is established for expenses incurred with the location, exploration, development and any other preparatory activity for exploitation of mineral resources. These expenses are recovered either through linear amortization[781] deductions throughout the duration of the contract or for the period of use, as the case may be..[782]

595. Holders of petroleum agreements and their subcontractors, who carry out petroleum operations in offshore perimeters which include areas with a water depth of over 100 m, are entitled to amortization of their tangible and intangible assets related to the petroleum operations, whose duration of use is limited by the duration of reserves, per each product unit, with an usability rate of 100%, depending on the exploitable resource, throughout the whole period of the petroleum agreement.[783] This provision was initially introduced by the Fiscal Code of 2003 and implemented by Decision 4/2004 of the Central Tax Commission.[784] It stated that with respect to the tax regime of expenses needed for the development and modernization of natural gas production, the 35% rate of the price per tons of extracted crude oil and delivered natural gas, calculated and collected according to Government Decision No. 168/1998[785] is to be reflected in reserve accounts, separated from current expenditures accounts. These sources are intended to cover the need for investments in the refining, transportation and distribution of petroleum, for geological

779. Article 4, letter b), point 9 of the Guidelines concerning deductions for research and development expenses during determination of taxable profits of 5 Jul. 2016.
780. Article 28, para. 3, letter c) of Fiscal Code.
781. Linear amortization is described in Art 28, para 6 of the Fiscal Code.
782. Article 28, para. 9 of Fiscal Code.
783. Article 26, paras. 2 and 3 of Fiscal Code.
784. Decision 4/05.07.2004, concerning the unitary application of certain provisions concerning the income/profit tax and excise taxes, published in the Official Gazette of Romania, Part I, No. 716/ 09.08.2004.
785. Government Decision No. 168/1998 setting the expense quotas needed for the development and modernization of crude oil and natural gas production, refining and petroleum transportation (hereinafter GD No. 168/1998), published in the Official Gazette of Romania, Part I, No. 128/ 27.03.1998, with subsequent amendments in 2000 and 2002.

research designed to find new reserves, for development of production, for retrofitting and modernizing existing equipment. The Decision emphasized that the amounts recorded in reserve accounts have the nature of subsidies for investments, both from an accounting and a taxation point of view. Therefore, throughout the amortization of the assets financed from this source these reserves are registered as revenues, according to the amounts recorded as expenditures.[786]

786. Point 1 of Decision 4/05.07.2004.

Chapter 3. Value Added Tax

596. According to Romanian law, the VAT is an indirect tax owed to the state's budget.[787] The basis for establishing the VAT is the value of the delivered goods or the performed service, of a taxable import or a taxable intra-community acquisition.[788] By goods, the law refers to tangible goods, movable and immovable, either by nature or by destination. For VAT purposes, electricity, thermal energy, natural gas, and other goods of the same nature, are deemed to be tangible movables. The law mentions that energy products are also subjected to excise tax with the sole exception of natural gas delivered through a natural gas system located on the EU territory or any other network connected to such system.[789]

597. The tax is owed by the taxable person who delivers goods or performs services, unless otherwise provided or who is the beneficiary of the services performed in Romania, delivered by a taxable person who is not a Romanian resident or not deemed registered for VAT purposes in Romania.[790] The law established who is deemed a taxable person for the purposes of VAT: any person who undertakes, in an independent manner and notwithstanding the location, economic activities mentioned by the law, disregarding the purpose and the result of such activities, or a public institution, when it is involved in delivery of gas, electricity, thermal energy or any other activities of the same nature.[791]

598. The economic activities referred to comprise in the activities of traders or service providers, including extractive activities or activities consisting in exploitation of tangible and intangible goods for the purpose of obtaining revenues on a continuous basis.[792]

599. With respect to the basis for VAT, regarding energy products, the law states that any intra-community acquisition of products subjected to excise tax is subjected to VAT.[793]

787. Article 265 of Fiscal Code.
788. Article 266, para. 1, point 5 of Fiscal Code.
789. Article 266, para. 1, point 6 and 7 of Fiscal Code. *See also* Art. 128, para. 12, e) defining 'non-transfers' as the delivery of gas through a natural gas system situated on the EU territory or any other network connected to such system, as well as electricity delivery and thermal energy delivery under Art. 132, para. 1, letters e) and f) of Fiscal Code defining the place of delivery. The place of delivery in such situations is deemed the place where the taxable trader has its economic activity registered office or a permanent office, for which the goods are delivered or the place where the buyer uses and effectively consumes the goods. A trader for the purposes of this article is a person who purchases gas, electricity and thermal energy for reselling it and whose own consumption of such products is negligible.
790. Article 307, paras 6 corroborated with Art 275, 278 and 316 of Fiscal Code.
791. Article 269, paras. 1 and7, letter b) of Fiscal Code.
792. Article 2, para. 2 of Fiscal Code.
793. Article 268, para 3, letter c) of Fiscal Code. According to Article 266, para 1, point 6 and 7, among products subjected to excise tax are electricity and natural gas as well as any other energy products.

600. The standard VAT rate is 19% starting with 1st of January 2017[794] and it is applied on the tax basis for all operations, unless they are exempted or subjected to reduced rate.[795] Energy products or services do not fall in any of the latter categories, so they are subjected to the standard rate.

601. With respect to exemptions from VAT, the law provides that delivery of goods which are to be admitted into the marine belt[796] in order to either be incorporated into drilling or production platforms, for the purpose of building, repairing, maintaining, modifying or reusing these platforms or for connecting these drilling or production platforms to the continent, or for supplying with fuel drilling or production platforms, are tax exempted.[797] Any taxable person has the right to deduct VAT related to its acquisitions if they are destined for the use of tax exempted goods such as those to be admitted in the marine belt to be incorporated in offshore platforms.[798]

794. Article 291, para.1, letter b) of Fiscal Code.
795. Article 291, para. 1 of Fiscal Code corroborated with point 36 of Guidelines to Fiscal Code.
796. The marine belt is defined in Art. 66, para 2 of Guidelines to Fiscal Code.
797. Article 295, para. 1, point 7 of Fiscal Code.
798. Article 297, para. 4, letter c) of Fiscal Code.

Chapter 4. Excise Tax

602. The level of harmonized excise taxes[799] is established in Annex 1, Positions 10–18, to Title VIII of the Fiscal Code covering excise and other special taxes. Starting with 2018, the level of excise tax applicable from 1 January of each year is the one set in columns 4–8 of Annex 1, updated in accordance with the provisions of the Fiscal Code.[800]

603. With respect to energy products and electricity, commercial usage means the usage of an undertaking – final consumer, who, repeatedly and in any place, delivers goods or services, notwithstanding the scope or the results of such economic activities. All activities of producers, traders or service providers, including mining activities, are deemed economic activities for the purpose of excise tax. Therefore, using electricity for public lighting is considered usage for commercial purpose.[801]

604. For the purposes of excise tax, production of excise tax products is any operation by which these products are produced, processed or transformed in any form,[802] including here the extraction operations of energy products (as defined by Article 355 of Fiscal Code[803]).[804] Energy products subjected to excise tax generally refer to energy products which can be used as fuels, gas and liquefied gas, coal, any additives or other product which is intended to be used as fuel or for thermal energy, all hydrocarbons (with the exception of peat) which are intended to be used as fuels or for thermal energy. If the consumption of such products takes place on the premises of a production facility of energy products it is not taxable if the consumption consists of products produced in that facility. However, if the consumptions occur in relation to other purposes than production, especially for fuelling vehicles, then the consumption is deemed to generate excise tax.[805]

605. Goods are subjected to excise tax either at the time of their production, including, where the case may be, their extraction within the community, or, at the time of their importation into the community. The excise tax becomes chargeable at the time of release for consumption in the Member State where the release takes places. However, moving energy products from the fiscal warehouse for the supply of ships and aircrafts which have a destination outside the community is not deemed

799. According to Art. 335, para. 2, letter c) harmonized excise taxes are (among others) special taxes levied directly or indirectly over the consumption of energy products and electricity.
800. Article 342, para 1 and 2 of Fiscal Code.
801. Point 4, paras 1 of Title VIII of Guidelines to Fiscal Code.
802. Exceptionally, additivation of or mixing energy products with products made of biomass under a suspension of excise tax can be conducted in fiscal warehouses.
803. Article 355, para. 1 of Fiscal Code provides a list of energy products and their codes while para. 3 provides a list of those products subjected to excise tax.
804. Article 336, point 19 of Fiscal Code.
805. Article 355, Para 8 of Fiscal Code corroborated with point 14, paras 7 of Title VIII of Guidelines to Fiscal Code.

release for consumption, being assimilated to an export operation. Similarly, fuelling ships and aircrafts which have a destination outside the community with energy products is also assimilated to an export operation. In all other cases where the tax was not chargeable for an energy product, whose initial destination was changed and which is offered for sale or used as fuel, the excise tax is chargeable from the date when the product is offered for sale or used as fuel. In the first case the burden of paying the tax lies with the seller, while in the second it lies with the user.[806]

606. According to the law, natural gas is subjected to excise tax, chargeable at delivery by distributors or authorized redistributors to final consumers. These undertakings must be registered with the central fiscal authority. However, if the natural gas is delivered in Romania by a distributor or redistributor from another Member State which is not registered in Romania, the excise tax is chargeable at delivery to the final consumer and is paid by an undertaking designated by the distributor or redistributor, which must be registered with the competent Romanian authority.[807]

607. With respect to coal, coke and lignite, the law states that they are also subjected to excise tax which is due at delivery by producers or traders who either engage in intra-community acquisitions or import such products. These undertakings must register with the competent authority.[808]

608. In what regards electricity, the law provides that it is subjected to excise tax, which is due upon supply to final consumers. However, electricity consumption will not be deemed to generate excise tax if it is used to maintain the capacity for production, transportation or distribution of electricity, within the limits established by NRAE. Similarly to situations concerning gas or coal, undertakings in the electricity area must register with the central fiscal authority. Another similarity refers to the case where electricity is supplied by a distributor or redistributor of another Member State, which is not registered in Romania. In this situation, the excise tax becomes chargeable at supply to final consumer and is paid by an undertaking designated by that distributor or redistributor, which must be registered.[809]

609. Harmonized excise taxes regarding products which are mandatorily subjected to excise tax at community level, covered by Directive 2008/118/CE, are calculated and paid in RON, according to the levels established in Annex 1 to Title VIII of the Fiscal Code. For energy products, the quantity is expressed in tons, 1,000 l or GJ, while for active electricity it is expressed in megawatts per hour.[810]

---

806. Article 340, para. 12 and of Fiscal Code.
807. Article 356 of Fiscal Code corroborated with point 15, paras. 4 and 5 of Title VIII of Guidelines to Fiscal Code.
808. Article 357 of Fiscal Code corroborated with point 16, paras 1) – 4) of Title VIII of Guidelines to Fiscal Code.
809. Article 358 of Fiscal Code, corroborated with point 17, paras 1) – 5)of Title VIII of Guidelines to Fiscal Code.
810. Annex 1 to Title VIII of Fiscal Code.

610. The law states that for energy products for which the level of excise tax is established for 1,000 l, the volume shall be measured at a temperature of 15°C.[811] In addition, it is specified that in case of diesel fuels to be used for agricultural purposes, a reduced excise tax is going to be applied.[812] A differentiated level of excise tax is also applied in the case of diesel fuel for engines used for the following purposes: road transport of merchandize with vehicles used exclusively for transport of goods and a maximum authorized gross weight of at least 7.5 tons or the transport of persons, with exception of local public transport, with an M2 or M3 category vehicle.[813] The differentiated excise tax is to be established by Government Decision.[814] Only registered[815] operators who meet the eligibility criteria[816] are entitled to receive the return.

611. The Fiscal Code established a series of exemptions from the obligation of paying excise tax for energy products and electricity. Thus, there is no excise tax on resulting heat or certain products (with codes NC 4401 and 4402). In addition, no excise tax is levied for the following usages of energy products and electricity: energy products are used for other purposes than as engine fuel or thermal energy fuel; dual usage of energy products (by dual usage the law understands those situations when the product is used both as thermal energy fuel and for other purposes, such as chemical reduction or electrolytic and metallurgical processes); electricity mainly used for chemical reduction and electrolytic or metallurgical processes; electricity when it represents more than 50% of the cost of a product; and for mineralogical processes.[817]

612. Other exemptions from excise tax payment with respect to energy products and electricity cover: energy products delivered to be used as fuel for aviation, other than private touristic aviation or for sailing in community or national waters, including fishing, other than sailing private pleasure crafts; electricity produced on the ships; energy products and electricity used for the production of electricity or for maintaining the electricity production capacity; energy products and electricity used for the combined production of electricity and thermal energy; energy products –

811. Article 342 para. 7 of Fiscal Code.
812. Article 342, para. 8 of Fiscal Code.
813. Article 342, para. 9 of Fiscal Code.
814. GD No. 537/2014 on conditions, terms and procedure of return as state aid of amounts representing the difference between the standard levels of excise tax and the differentiated levels of excise tax for disel fuels used as engine fuels, was published in the Official Gazette of Romania, Part I, No. 512/09.07.2014, subsequently amended in 2014, 2015, and 2016 (hereinafter GD No. 537/2014). The current rates were established by GD No. 961/2016 for the modification and completion of GD No. 537/2014, published in the Official Gazette of Romania, Part I, No. 1037/22.12.2016.
815. Article 8, para. 1 of Decision No. 537/2014.
816. Articles 4 and 5 of Decision No. 537/2014.
817. Article 394 of Fiscal Code corroborated with Point 73 of Title VIII of the Guidelines to Fiscal Code. By mineralogical processes, the law refers to processes qualified under CAEN, section C, division 23 of Regulation CE No. 1893/2006, as specified at Point 71 of Tivle VIII of the Guidelines to Fiscal Code.

natural gas, coal and solid fuels – used by households and/or charitable organizations; fuels for engines used in production, development, testing and maintenance of aircrafts and ships; fuels for engines used in dredging operations in navigable watercourses and ports; mineral oils injected into blast furnaces or other industrial equipment for chemical reduction purposes as an additive to coke used as the main fuel; energy products entering Romania from a third country contained in the standard tank of a motor vehicle; energy products released from the state's reserve as humanitarian aid; any energy product used as fuel for thermal energy of hospitals, hospices, asylums, orphanages and other social assistance institutions, educational institutions and warship places; energy products obtained from biomass or biofuels; and energy produced from renewables.[818]

613. Excise taxes are a source of revenue for the state's budget, and, as mentioned at the beginning of this chapter, a very important one. The deadline for payment is the twenty-fifth of the following month to the one in which the taxes become due, unless otherwise stated in the law. However, authorized suppliers of electricity or natural gas, must pay until the twenty-fifth of the following month to the one in which the final consumer was invoiced.[819]

614. Any delay in payment of excise taxes which is longer than fifteen days may attract the suspension of the authorization held by the undertaking until payment of all arrears.[820] The law also provides for a large list of misdemeanours[821] and criminal offences related to production or distribution of excise goods in violation of the legal provisions.[822]

818. Article 206^{60} of Fiscal Code, corroborated with point 113. 19 of Guidelines to Fiscal Code.
819. Article 345, paras 1 and 2 of Fiscal Code.
820. Article 433 of Fiscal Code corroborated with point 159 of Title VIII of Guidelines to Fiscal Code.
821. Article 449 of Fiscal Code.
822. Article 452 of Fiscal Code.

Chapter 5. Special Taxes

§1. Tax on Domestic Oil

615. Until 2015, according to Romanian law, for oil from domestic production all authorized undertakings had to pay a tax to the state's budget, at the moment of delivery. The amount was set by the Fiscal Code of 2003 at RON 18.95 per ton (approximately EUR 4), and was supposed to be updated yearly. The tax was applied on delivered quantities and was chargeable at delivery date, until the twenty-fifth of the following month to that when the tax became due.[823]

616. However, the tax was not maintained in the Fiscal Code adopted in 2015, which is currently in force.

§2. Special Buildings Tax or 'The Pole Tax'

617. In 2014, the Romanian government introduced a tax on special buildings which did not fit into the taxable basis such as: poles, ramps, concrete platforms other than production and office buildings, which were already taxed. The initial amount was 1.5% of the building's gross accounting value and was a revenue source for the state's budget.

618. Among the most affected undertakings by the pole tax were those acting in the energy sector, conducting oil and gas extraction and mining operations, given the high value of special buildings held by such undertakings.

619. Despite heavy criticism from the energy business, which was heavily affected by the measure, the tax brought three times more revenues than it was estimated: RON 1.5 billion (around EUR 340 million).

620. However, the tax was reduced to 1% in 2015.[824] Subsequent discussions concerning the modification of the basis for calculation of the pole tax were silenced by the Government's promise to remove the tax from 2016. Ultimately, the tax was removed from the Fiscal Code of 2015.

§3. Other Special Taxes outside the Fiscal Code: Tax on Natural Monopolies in the Electricity and Natural Gas Sector

621. Considering that monopolies in general can lead to:

- Maximizing profits in the absence of demand elasticity, a situation which occurs naturally in the electricity sector and the gas.

823. Article 215 of Fiscal Code of 2003 corroborated with point 32 of Guidelines to Fiscal Code 2003.
824. Article 296^{35} of Fiscal Code and the applicable dispositions of Guidelines to the Fiscal Code.

– Exercising market power by raising prices far above marginal cost, without this leading to a loss of customers, even in the case of regulations issued by regulatory authorities in the field.

The Government decided to impose a special tax on natural monopolies in the electricity and natural gas sector, which would constitute a revenue to the state's budget.[825]

622. The special tax is owed by electricity and natural gas transport operators, authorized by NRAE and by undertakings who distribute electricity and natural gas on the basis of concession agreement concluded with the Ministry of Economy and Commerce or with local authorities.[826]

623. The tax is levied for each megawatt per hour for which services of transport and distribution of electricity and gas are invoiced, until the end of 2018.[827]

§4. The Quota for Development and Modernization of Oil and Gas Production

624. As already stated, the government established a quota of 35% of the price per ton of extracted oil and delivered natural gas, for development and modernization of production, refining and transportation.[828] In addition, the quota is subtracted from the tariff for storage of natural gas; the transportation tariff; or, the distribution tariff, as invoiced by Romanian or foreign authorized or licensed legal persons to operate an underground storage system, a transport system or, respectively, the distribution and commercialization activity of natural gas. The quotas are included in the delivery prices,[829] and may be calculated, collected and included by undertakings only throughout the capitalization and collection of the production.[830]

825. Article 1 of Government Ordinance No. 5/2013 (GO No. 5/2013) on establishing special measures of taxation of natural monopoly activities in the electricity and natural gas sector, published in the Official Gazette of Romania, Part I, No. 52/2013, subsequently amended in 2014, 2015, 2016 and 2017.
826. Article 2, paras 1 and 2 of GO No. 5/2013.
827. Articles 3 and 6 of GO No. 5/2013, as amended by GO No. 90/2017 concerning certain fiscal-budgetary measures, the modification and completion of certain normatives acts and the extension of several deadlines, published in the Official Gazette of Romania, Part I, No. 973/07.12.2017.
828. Article 1 of GD No. 168/1998.
829. Article 2 of GD No. 168/1998.
830. Article 3 of GD No. 168/1998.

Chapter 6. Local Taxes

§1. Land Tax

625. For lands belonging to the private or public property of the state or administrative units, which are given in concession, rented, or given in administration or use, a land tax is set in the burden of concession holders, tenants with rights of administration or use, as the case may be. The land tax is owed to the local budget of the administrative unit where the land is located.[831]

626. However, the land tax is not owed for any degraded or polluted land, included in an rehabilitation perimeter, for the period of the rehabilitation or for lands which are used as legally designed protection areas as well as for lands used for underground exploitation, which are so determined by a decision of the local council, insofar as it does not affect the usage of the soil surface.[832]

§2. Tax for Issuance of Permits, Approvals or Authorizations

627. By law any person who must obtain a certificate, approval or any other authorization must pay a tax to the specialized department of the local public authority, before the issuance of the needed certificate, approval or authorization. Undertakings acting in the energy sector are also bound by this obligation.[833]

628. In order to be allowed to build or install equipment needed for their operations, undertakings must obtain first an urban certificate, an authorization for construction and any other related approval. For each, they must pay the amount of tax set by the Fiscal Code. For example, the tax for the issuance of the permit for drilling or excavation needed for geotechnical studies, surveys, mines, ballast, oil and gas wells and any other operations is calculated by multiplying the number of square metres of land affected by drilling or excavation with a value set by the local council up to RON 15 (less than EUR 4).[834] The tax for issuance of the authorization for construction concerning energy undertakings is equal with 1% of the authorized value of the construction works, including the related equipment.[835]

629. The tax for authorization of total or partial dismantling of a building is set to 0.1% of the taxable value of the building,[836] while the tax for authorization of

831. Article 463, paras 2 and 3 of Fiscal Code.
832. Article 464, letters i) and j) of Fiscal Code, corroborated with points 72, respectively 73 of Title IX of Guidelines to Fiscal Code.
833. Article 473 of Fiscal Code, corroborated with point 138 of Title IX of Guidelines to Fiscal Code.
834. Article 474, paras 10 of Fiscal Code, corroborated with point 144 of Title IX of Guidelines to Fiscal Code.
835. Article 474, para. 6 of Fiscal Code corroborated with point 140 of Guidelines to Fiscal Code.
836. Article 474, para. 9 of Fiscal Code corroborated with point 142–143 of Title IX of Guidelines to Fiscal Code.

works on fittings and connections to public water, sewer, gas, thermal energy and electricity is set by the local council up to RON 13 (about EUR 3) for each connection.[837]

§3. Other Special Taxes

630. For the functioning of public local services which were created to the benefit of natural and legal persons, local or county councils are entitled to adopt special taxes. The Fiscal Code of 2003 provided that local or county councils are not entitled to impose any other special taxes on concession holders from public utilities sector (natural gas, water or electricity) for the presence on the public or private domain of administrative units of networks for transport and distribution of electricity and natural gas.[838] However, the aforementioned provision is not found in the Fiscal Code of 2015 that is currently into force. On the contrary, one provision states that local councils may establish taxes for owning and using equipment and machinery destined to generate profits, which use the local public infrastructure, as well as taxes for activities that have an impact on the environment.[839]

837. Article 474, para. 15 of Fiscal Code corroborated with point 145, paras. 1 and 3 of Title IX of Guidelines to Fiscal Code.
838. Article 282, paras 1 and 4 of Fiscal Code of 2003 corroborated with point 195, para. 3 of Guidelines to Fiscal Code of 2003.
839. Article 486, para 2 of Fiscal Code.

Part VII. Interaction Between Energy Law and Competition Law

631. The EU policy regarding energy is based on three main objectives: durability, competitiveness and safety in energy supply. In order to ensure competitiveness an internal functional energy market is needed. At EU level several measures have been established to be implemented by the European Commission together with the Member States. Thus, with respect to the internal market, the EU energy policy seeks: to ensure the functioning of energy markets with observance of fair competition rules and to promote the interconnectivity of energy networks.

632. The energy packages adopted by the EU between 1996 and 2009 followed up on the aforementioned objectives and regarded: the separation of energy production and supply from the transport networks; the consolidation of consumers' rights; the increase of investments in cross-border infrastructure; a bigger market transparency with respect to the functioning of networks and supply; and a more effective regulation at national level together with a greater solidarity among EU states. The EU has also regulated cross-border exchanges of electricity, a wholesale, transparent and functional market, a European network of transport and system operators for electricity, the foundation of an independent and specialized organism for cooperation between national energy authorities and monitoring of wholesale electricity market. Last but not least, measures and objectives for common framework regarding the promotion of renewable energy at EU level have been implemented. The Romanian legislator has transposed most of these measures in the national legislation but the secondary legislation is still undergoing a process of correlation with the EU requirements.

Chapter 1. General Introduction

633. However, unlike other sectors, liberalizing the Romanian energy market proved to be a more difficult and sensitive topic, at least from a political standpoint. Since Romania does not depend as much as other EU Member States on energy imports – the entire demand for electricity and most of the demand for natural gas can be covered by internal production – governments have been able to maintain lower tariffs for energy.[840] Thus, for a long time there was no will to completely liberalize the market, for that would have affected the public perception of the party in government for future elections. Currently, all legal steps necessary for liberalization have been taken, but as it was shown, regulated markets continue to co-exist along with competitive markets, which are limited to large industrial users, which causes significant market distortions. Therefore Romania now has a roadmap for abandoning all price regulation: in 2017 for electricity and 2018 for gas.[841]

634. Given the situation described above it is not surprising that in 2013, according to a statement of the Competition Council's president, the largest number of issues regarding competition are raised by the energy sector. The reasons were, on the one hand, that there was still a large number of state-owned companies, and, on the other hand, that the market was heavily regulated. Thus, since the energy sector is still undergoing a process of liberalization, it continues to raise competition issues and requires the close scrutiny of the competent authority.[842]

635. In 2017, according to the Competition Council's Annual Report 2016, the energy sector remains the top priority. The Competition Council was planning to finalize a sectorial investigation concerning natural gas and two ongoing investigations concerning fake tenders for supply of electricity meters and other equipment ancillary to measurement of electricity consumption, respectively acquisition of natural gas connectors, modernization and maintenance of related infrastructure.[843] Additionally, the sectorial investigation regarding the energy sector started in 2009, has produced a series of results and conclusions which were made public 2015.[844]

840. OECD, *Romania. Competition Law and Policy. A Peer Review* 76 (2014), available online at http://www.oecd.org/daf/competition/Romania-Competition-Law-Policy-2014-EN.pdf, last visited on 23 Mar. 2018.

841. OECD, *Romania. Competition Law and Policy. A Peer Revie* 77 (2014), available online at http://www.oecd.org/daf/competition/Romania-Competition-Law-Policy-2014-EN.pdf, last visited on 23 Mar. 2018.

842. Chiritoiu: Piata de energie, cele mai multe probleme concurentiale. Auto si telecom cele mai putine, 7 Nov. 2013. Available online, in Romanian only, at: http://www.economica.net/chiritoiu-piata-de-energie-cele-mai-multe-probleme-concurentiale-telecom-ul-si-auto-cele-mai-putine_66072.html, last accessed 23 Mar. 2018.

843. Consiliul Concurentei – Raport Anual 2016, pp. 7–8, available on line, in Romanian, at: http://www.consiliulconcurentei.ro/uploads/docs/items/bucket12/id12290/raport_cc_2016-final.pdf, last visited on 23 Mar. 2018.

844. Consiliul Concurentei, Directia Industrie si Energie – Raport privind rezultatele investigatiei sectoriale pe piata energiei electrice din Romania, available online, in Romanian only, at http://www.consiliulconcurentei.ro/uploads/docs/items/bucket11/id11891/raport_privind_rezultatele_investigatiei_sectoriale_pe_piata_energiei_electrice.pdf, last visited 23 Mar. 2018.

§1. Romanian Competition Law

636. The general framework covering competition is Competition Law No. 21/1996[845] which has the purpose of protecting, preserving and stimulating competition and a normal competition environment for the purpose of promoting consumers' interests. The law thus covers actions and deeds, which restrict, preclude or distort competition perpetrated by undertakings or groups of undertakings – both natural and legal persons – notwithstanding whether they are national or foreign. By undertaking, the law understands any undertaking engaged in economic activities consisting in offering of goods or services on a given market, irrespective of its legal status and financing method, as defined by the EU case law.[846]

§2. Main Violations

637. Romanian competition law bans any agreements between undertakings, decisions of shareholders and concerted practices which have as object or effect hindering, restraining or distorting of competition on the Romanian market, in part or as a whole, especially those which: indirectly or directly set sell and purchase prices or any other conditions of the transaction; limit or control the production, trade, technical development or investment; divide markets or supply sources; apply, in relation to business partners, unequal conditions for equivalent performances, thus causing to some of them a competitive disadvantage; condition the conclusion of contracts of acceptance of other ancillary performances which do not have any connection with the objects of these contracts; consist of participating in concert to fake tenders or any other public calls for offers; and eliminate other competitors from the market, limit or impeach their access to the market and free competition among undertakings.[847]

638. Another main violation consists in abusing a dominant position held by an undertaking, or a group of undertakings, on the Romanian market (or a substantial part of it) which results in any of the actions described in Article 5.[848] Furthermore, Competition Law bans any actions or inactions of public administrative authorities or bodies, either central, or local, and of any subordinated entities, which restrict,

---

845. Competition Law No. 21/1996 was republished in the Official Gazette of Romania, Part I, No. 153/29.02.2016, and was subsequently amended in 2017.

846. Articles 1 and 2, para. 1, letter a) and para. 2 of Competition Law. Art 2, para. 2 was amended by GEO No. 39/2017 regarding compensation actions arising from breaches of legal provisions in the area of competition and for the modification and completion of Competition Law No. 21/1996, published in the Official Gazette of Romania, Part I, No. 422/08.06.2017.

847. Article 5, para. 1 of Competition Law. The burden of proof for any alleged violation of these provisions belongs to the Competition Council. Details concerning the application of Art. 5 are contained in the Guidelines regarding conditions, terms and procedure for accepting and evaluating commitments, in the case of anti-competitive practices of 28 Dec. 2010, published in the Official Gazette of Romania, Part I, No. 11/05.01.2011, subsequently amended in 2011, 2012 and 2014.

848. Article 6, para. 1 of Competition Law.

impeach or distort competition such as: limiting the freedom of commerce or company autonomy and setting up discriminatory conditions for the activity of undertakings. In such cases, the Competition Council may impose on breaching authorities an obligation to restore the competitive environment within a certain deadline. Failure to comply with the Council's decision may result in judicial action in administrative courts.[849]

849. Article 8 of Competition Law.

Chapter 2. The Competent Authority

639. The enforcement of Competition Law lies within the competences of the Competition Council, as national autonomous authority.[850]

§1. The Competition Council

640. The Competition Council seeks the protection, maintenance and stimulation of competition and of a normal competitive environment, in order to promote consumers' interests.[851] The Competition Council applies the dispositions of Articles 101 and 102 of the Treaty for the Functioning of the EU (TFEU) according to the provisions of Regulation 1/2003 concerning the Guidelines for application of competition rules provided by Articles 81 and 82 of the treaty, with subsequent amendments, in all cases where actions and deeds of undertakings and groups of undertakings could affect commerce between EU Member States.[852]

§2. Main Powers and Duties

641. By law, the Competition Council has the following main duties: investigates the application of the provisions of Articles 5, 6, 7, 8, 13 of the Competition Law and of Articles 101 and 102 of TFEU; takes legal measures in case of violations of the aforementioned articles; accepts settlements and imposes interim measures; takes legal measures with regard to economic concentrations; monitors and enforces its decisions; investigates on its own initiative, specific economic sectors or specific types of agreements within different sectors, when the rigidity of prices or other circumstances suggest the possibility of a restriction or distortion of competition on the market; informs the government on the existence of a monopoly situation and proposes measures for the remediation of the disruptions found; petitions the courts in cases where the judiciary has jurisdiction; issues approvals for draft normative acts which may have an anti-competitive impact and proposes amendments of acts which have such an effect; and ensures the application of the legal provisions concerning the area of state aid and unfair competition.[853]

642. Among the legal powers granted to the Competition Council in order to perform its duties, one may find the possibility to: perform preliminary examinations during which it may solicit all necessary documents and information; issue

850. Beside the provisions of the Competition Law, the functioning of the Competition Council functions in accordance with the Regulation concerning the organization and functioning of the Competition Council of 29 Jun. 2017 (hereinafter ROFCC), published in the Official Gazette of Romania, Part I, No. 601/26.07.2017, with the subsequent amendments of 2018.

851. Article 2 of ROFCC.

852. Article 3 of Competition Law.

853. For a full list of duties *see* Art 25, para. 1 of Competition Law corroborated with the provisions of Arts 9–30 of the ROFCC.

guidelines and instructions, orders, decisions and approvals, make recommendations and elaborate reports concerning the application of the Competition Law; and make 'good practice' recommendations in diverse economic sectors or concerning diverse general aspects of applying the legislation concerning competition, by taking into consideration both the case law of the national and the EU courts or the practice of the European Commission.[854]

643. The decisions issued by the Council are individual administrative acts, which establish the violation of the Competition Law and the concomitant sanctions, the measures needed to restore the competitive environment, grant access to confidential information, settle the complaints raised under the law as well as the requests and notifications concerning mergers.[855] Given the administrative character of the decisions issued by the Competition Council, any complaint against it attracts the jurisdiction of the administrative courts as provided by Law No. 554/2004 concerning administrative litigation.

854. Article 26, paras 1–3 of Competition Law corroborated with Arts 9–11 of ROFCC. The procedure is elaborated in Arts 32–49 of Competition Law.
855. Article 26, para. 4 of Competition Law.

Chapter 3. The Electricity Sector and Electricity Market

644. The electricity sector is very peculiar. The constant change in demand for electricity and the necessity of producing at the same time with consuming requires a high degree of coordination between production, transportation and distribution activities across the entire country. Thus, the law imposes a requirement for all participants on the market to function on a basis of a license and comply with all applicable legal provisions.

645. Electricity is a unique *merchandize*, with *special features* which have a significant impact on the structure and competition on the electricity market. The most important one derives from the fact that electricity cannot be stored in conditions of economic efficiency which leads to the need to constantly balance production with consumption. Electricity is produced in power plants and delivered through networks, which constitute natural monopolies. Therefore, the electricity market can only function if supported by the system's infrastructure. Additionally, production costs vary significantly depending on the technologies used. Electricity may be imported and exported, in higher or smaller quantities, depending on the interconnectivity capacity between the neighbouring states. However, on short term, the elasticity of demand for electricity is very low, an increase of price having a very limited influence on consumption.[856]

646. In the energy sector, all activities of production and supply of electricity take place on a competitive basis, while transmission and distribution of electricity through electricity networks constitutes natural monopoly and takes place under a regulated regime.[857]

647. In 2014, state-owned companies still dominated the *production*, accounting for 80%–90% of the entire power generation, while private owned companies (mostly from gas-fired plant and renewables) account for about 13%. Even so, the Competition Council considers that the production companies are individually managed and in competition with each other, despite the fact that others have described the situation as being a 'soft cartel'.[858]

856. Consiliul Concurentei, Directia Industrie si Energie – Raport privind rezultatele investigatiei sectoriale pe piata energiei electrice din Romania, available online, in Romanian only, at http://www.consiliulconcurentei.ro/uploads/docs/items/bucket11/id11891/raport_privind_rezultatele_investigatiei_sectoriale_pe_piata_energiei_electrice.pdf, last visited 23 Mar. 2018.

857. Raport privind rezultatele preliminare ale investigatiei sectoriale pe piata energiei electrice din Romania, p. 19, available in Romanian only at *http://media.hotnews.ro/media_server1/document-2015-03-3-19534917-0-raport-investigatie-utila-energie-electrica.pdf*, and Raport privind rezultatele investigatiei sectoriale pe piata energiei electrice din Romania, p. 5, available online, in Romanian only, at http://www.consiliulconcurentei.ro/uploads/docs/items/bucket11/id11891/raport_privind_rezultatele_investigatiei_sectoriale_pe_piata_energiei_electrice.pdf, last visited on 23 Mar. 2018.

858. OECD, *Romania. Competition Law and Policy. A Peer Review* 77–78 (2014), available online at http://www.oecd.org/daf/competition/Romania-Competition-Law-Policy-2014-EN.pdf, last visited on 23 Mar. 2018.

648. The *transmission* network is operated by a state-owned independent system operator, Transelectrica, while the market operator (OPCOM) administers the wholesale electricity and green certificates market. As shown in the first chapter, they both are under the supervision of NRAE. As the next subsection reveals, OPCOM was responsible for several infringements of competition law.

649. The *distribution* network is owned by distribution companies, which enjoy a local monopoly based on concession. Most of the distribution companies have, in fact, been privatized and are now subsidiaries of large European energy holdings. Both wholesale and retail *supply* is now opened to competition. However, cross-border trade is still underdeveloped due to a limited interconnector capacity with neighbouring countries.[859]

650. As it was previously stated, the Romanian electricity market has two components: the regulated market, which functions on the basis of regulated contracts and regulated tariffs, and the competitive market, which functions on the basis of commercial contracts, with negotiated and prices and quantities, concluded between producers, suppliers and/or clients. On the latter market, the prices are freely established on the basis of supply and demand, with NRAE regulating the functioning of the market and the Competition Council supervising it. Sales of electricity may be either wholesale, or retail.

§1. The Wholesale Electricity Market Operator (OPCOM) and Restriction of Competition

651. Starting with 2012, when the new Electricity Law came into force, all wholesale transactions take place through the trading instruments provided by the Electricity Market Operator (OPCOM). As a result, for this market segment, the possibility of direct negotiation between parties entering into wholesale contracts has been completely eliminated. At the same time, electricity contracts cannot be intermediated by other trading platforms. This aspect raised a lot of concerns and criticism from business organizations[860] or other trading platforms (such as the Romanian Commodities Exchange),[861] both accusing the new law of instituting a legal monopoly in the area of trading on regulated wholesale electricity market.

652. OPCOM, jointly and severally with its parent company, Transelectrica, has also become the first energy-related institutions in Romania to be sanctioned by the European Commission for abuse of dominant position in the Romanian market for

859. *Ibid.*, at 78–79.
860. AOAR: Noua lege a energiei permite instaurarea monopolului, 19 Sept. 2012, available online in Romanian only at: http://www.business24.ro/energie/stiri-energie/aoar-noua-lege-a-anergiei-permite-instaurarea-monopolului-1519092, last visited 23 Mar. 2018.
861. Alexandru Moldovan – De la monopol la concurenta, 2 Apr. 2012, available online in Romanian only at: http://energy-center.ro/piata-energiei-din-romania/de-la-monopol-la-concurenta/, last visited 23 Mar. 2018.

facilitating electricity spot trading and breach of EU anti-trust law. The Commission concluded that for a period of over five years OPCOM has discriminated against EU located electricity traders from outside Romania by creating an artificial entry barrier. Basically, between 2008 and 2013 OPCOM has imposed on members of spot electricity markets an obligation to register in Romania for VAT taxes although many of them were registered for VAT in the other EU Member States. Therefore, EU traders could not enter the Romanian wholesale market unless they had a permanent office in Romania, which resulted in additional costs and disadvantages for EU traders in comparison with their Romanian counterparts. Since discrimination on grounds of nationality or place of establishment is against the basic principle of the single market, OPCOM was held liable for infringement and sanctioned with a fine of over EUR 1 million.[862]

653. The Romanian Competition Council's Preliminary Report on Electricity Markets reveals additional issues related to OPCOM.[863] Thus, the Council stated that the provisions of Electricity and Natural Gas Law No. 123/2012, which require the traders to conclude wholesale agreements only on the centralized markets could be interpreted as an interdiction for Romanian electricity producers of exporting energy. However, the Council interpreted the legal provisions as allowing the producers to sell trade directly for exporting purposes and thus found no violation of anti-trust rules.[864]

§2. The Regulated Retail Energy Market and Restriction to Competition

654. The major source of distortions to effective competition remains the large regulated retail market. In Romania, private consumers and small companies are still the beneficiaries of regulated low prices, established by NRAE. The legal option of switching to competitive suppliers does not appear to be attractive, especially since the transition to the fully competitive market will go on until the end of 2017. Obviously, this transition will create major challenges for both the energy sector regulator and the Competition Council. Any increase in prices for final consumers will be met with dissatisfaction, which might reflect on votes and affect the political will to try to maintain price regulation.[865]

---

862. http://europa.eu/rapid/press-release_IP-14-214_en.htm, last visited on 23 Mar. 2018.
863. Consiliul Concurentei, Directia Industrie si Energie – Raport privind rezultatele investigatiei sectoriale pe piata energiei electrice din Romania, pp. 8–9, 14–15, available online, in Romanian only, at http://www.consiliulconcurentei.ro/uploads/docs/items/bucket11/id11891/raport_privind_rezultatele_investigatiei_sectoriale_pe_piata_energiei_electrice.pdf, last visited 23 Mar. 2018.
864. *Ibid.*, at 8.
865. OECD, *Romania. Competition Law and Policy. A Peer Review* 80 (2014), available online at http://www.oecd.org/daf/competition/Romania-Competition-Law-Policy-2014-EN.pdf, last visited on 23 Mar. 2018.

§3. The Electricity Market: Risks and Council's Recommendations

655. The analysis of the Romanian legislation in the field of electricity led the Competition Council to one of the most important observations and recommendations concerning the normative framework. The Council stated that the national legislation does not consider specific notions related to competition, although they are regulated at community level. Thus, the national legislation does not take into account notions such as: 'group of undertakings', 'parent company', 'intra-group transactions', 'single economic unit' even though these notions are implemented in Electricity and Gas Law No. 123/2012, which leads to distortions in the functioning of the competition mechanisms. The Council also observed that intra-group transactions are in fact performed through the platforms provided by OPCOM. This leads to the situation where the parent company of a group is a competitor to another company in which it holds controlling shares, although the former is entitled to coordinate the market behaviour of the latter. In conclusion, the Competition Council recommends to correlate the national legislation with community legislation by implementing the regulated community notions into the national legislation.[866]

656. As part of its preventive duties and based on its own researches, the Competition Council has stressed its concern with the fact that a uncorrelated development of investments, without ensuring a legislative stability which would allow a certain degree of predictability, might lead to an increase of electricity costs borne by final consumers.[867]

657. It also recommended either the removal of the restriction to trade outside the centralized markets, or to create a clear functioning mechanism which would reduce the barriers related to export sales by Romanian electricity producers. In direct connection with OPCOM's legal monopoly, the Competition Council recommended to open the electricity market to other intermediaries, in order to remove any possibility of distorting the competitiveness of the electricity market.[868]

658. The Preliminary Conclusions of the Report concerning the investigation conducted on the Romanian electricity market revealed the suspicion that some of the recent legislative changes[869] might have led to a dominant position of state-owned companies on the Romanian market for production and trade of electricity. The concern of the Council was raised by the fact that the administrative powers of

866. Consiliul Concurentei, Directia Industrie si Energie – Raport privind rezultatele investigatiei sectoriale pe piata energiei electrice din Romania, p. 19–22, available online, in Romanian only, at http://www.consiliulconcurentei.ro/uploads/docs/items/bucket11/id11891/raport_privind_rezultatele_investigatiei_sectoriale_pe_piata_energiei_electrice.pdf, last visited 23 Mar. 2018.

867. Idem at 7.

868. *Ibid.*, at 19.

869. *See* Government Decision (GD) No. 429/2013, published in the Official Gazette of Romania, Part I, No. 419/10.07.2013, (repealed by GD No. 42/2015), by which the controlling shareholder enjoyed, along with the right which derive from this status, the right to administer the companies in its portfolio, which was equivalent to a control right held by a parent company in a group of companies.

the controlling shareholder have increased, which affects negatively the independence of decision making in these companies.[870] Although the normative act considered by the Competition Council was repealed, the replacing document and the subsequent normative changes[871] maintained the criticized provision, thus, the Council's concern is still current.

659. The Competition Council observed also the lack of objective criteria to justify the difference in tariffs and the segmentation of the markets for production and trading of electricity in Romania by applying different tariffs to different zones. The risk identified by the competent authority is that such differences cause a distortion in competition without the possibility to identify any benefits for consumers.[872] In addition, the Competition Council emphasized the necessity to eliminate restrictions on concluding commercial agreements for exporting electricity. A preliminary analysis seemed to indicate that the provisions of the Electricity Law that oblige market participants to conclude *wholesale* transactions solely on the centralized markets administered by OPCOM, could be interpreted as an interdiction to conclude export agreements for Romanian electricity producers. Since an obligation to offer the entire production solely on the national markets would constitute a conspicuous violation of Union law, the Competition Council took the view that Electricity Law cannot be interpreted in a manner contrary to the principles of EU law. Given the ambiguous wording of the Electricity Law, the Competition Council recommended NRAE to solve the matter by issuing regulations and by urging the regulatory agency to identify trading platforms at EU level that meet the transparency and non-discriminatory requirements of the law combined with creation of a possibility to make export offers on OPCOM platforms, by introducing new transaction instruments.[873]

870. Consiliul Concurentei, Directia Industrie si Energie – Raport privind rezultatele preliminare ale investigatiei sectoriale pe piata energiei electrice din Romania, pp. 153–154, available online, in Romanian only, at http://www.consiliulconcurentei.ro/uploads/docs/items/id10016/raport_investigatie_utila_energie_electrica_07012014-forma_neconfidentiala-pub.pdf, last visited 23 Mar. 2018.

871. *See* Art. 3, point 6 of GD No. 42/2015, concerning the organization and functioning of the Ministry of Energy, Small and Medium Enterprises and the Business Environment, published in the Official Gazette of Romania, Part I, No. 75/28.01.2015 (repealed by GD No. 980/2015) and Art. 3, point 6 of GD No. 980/2015, published in the Official Gazette of Romania, Part I, No. 952/22.12.2015.

872. Consiliul Concurentei, Directia Industrie si Energie – Raport privind rezultatele investigatiei sectoriale pe piata energiei electrice din Romania, pp. 21, available online, in Romanian only, at http://www.consiliulconcurentei.ro/uploads/docs/items/bucket11/id11891/raport_privind_rezultatele_investigatiei_sectoriale_pe_piata_energiei_electrice.pdf, last visited 23 Mar. 2018.

873. Consiliul Concurentei, Directia Industrie si Energie – Raport privind rezultatele investigatiei sectoriale pe piata energiei electrice din Romania, pp. 7–8, available online, in Romanian only, at http://www.consiliulconcurentei.ro/uploads/docs/items/bucket11/id11891/raport_privind_rezultatele_investigatiei_sectoriale_pe_piata_energiei_electrice.pdf, last visited on 23 Mar. 2018.

§4. Electricity Market: Concluded and Pending Investigations

660. The competent authority reported also a series of concluded and pending investigations related to the electricity market. These investigations may help understand which issues connected to anti-trust law arise in Romania.

661. For example, the Competition Council scrutinized the long-term bilateral contracts entered into by one of the major electricity producers in Romania (SC Hidroelectrica SA) with respect to their: duration, market position of parties involved and the contractual terms, in order to assess whether the existence and performance has significantly influenced the electricity market at national level. The preliminary analysis has revealed that possible breaches of Article 5, paragraph 1 of Competition Law might have occurred and an investigation was open.[874] On 24.12.2015, the Competition Council issued Decision No. 82, by which it found that Article 5 was breached and the conclusion and performance of the contracts in question blocked the access to the market for other suppliers and producers of electricity and, thus, hardened the development of the market, in a period when the sector was under liberalization. In addition, it found that many of the contractual partners of Hidroelectrica have commonly exercised their purchase power and coordinated their competitive behaviour, thus setting the purchase conditions, including the price.[875] These contracts are also under investigation by the European Commission given that low prices for electricity for major industrial customers may be deemed state aid.[876]

662. Another pending investigation relates to possible violations by the NRAE of Article 8, paragraph 1, of Competition Law.[877] After analysing the new Guidelines for setting prices for electricity sold by producers on the basis of regulated contracts and quantities originating in regulated contracts between producers and end suppliers,[878] the Council concluded that the 'energy basket' for the regulated market is susceptible of distorting competition between electricity producers acting on the competitive segment. In other words, the Council established that the

874. Consiliul Concurentei, Directia Industrie si Energie – Raport privind rezultatele preliminare ale investigatiei sectoriale pe piata energiei electrice din Romania, pp. 158–159, available online, in Romanian only, at http://www.consiliulconcurentei.ro/uploads/docs/items/id10016/raport_investigatie_utila_energie_electrica_07012014-forma_neconfidentiala-pub.pdf, last visited 23 Mar. 2018.

875. Raport privind rezultatele preliminare ale investigatiei sectoriale pe piata energiei electrice din Romania, p. 17, available online, in Romanian only, at http://www.consiliulconcurentei.ro/uploads/docs/items/bucket11/id11891/raport_privind_rezultatele_investigatiei_sectoriale_pe_piata_energiei_electrice.pdf, last visited on 23 Mar. 2018.

876. OECD, *Romania. Competition Law and Policy. A Peer Review* 79 (2014), available online at http://www.oecd.org/daf/competition/Romania-Competition-Law-Policy-2014-EN.pdf, last visited on 23 Mar. 2018. *See also* the European Commission's press release at: http://europa.eu/rapid/press-release_IP-12-397_en.htm, last visited on 23 Mar. 2018.

877. *See* Order of the President of the Competition Council No. 143/14.04.2014.

878. *See* Order No. 83/20.11.2013, published in the Official Gazette of Romania, Part I, No. 732/28.11.2013.

regulation indirectly establishes the market position of each producer on the competitive market which is likely to be discriminatory and create distortions on the producing and trading electricity market.[879]

879. Consiliul Concurentei, Directia Industrie si Energie – Raport privind rezultatele preliminare ale investigatiei sectoriale pe piata energiei electrice din Romania, pp. 160–161, available online, in Romanian only, at http://www.consiliulconcurentei.ro/uploads/docs/items/id10016/raport_investigatie_utila_energie_electrica_07012014-forma_neconfidentiala-pub.pdf, last visited 23 Mar. 2018. *See also* Order of the President of the Competition Council No 143/14.04.2014.

Chapter 4. The Gas Market

663. The gas market is in a similar situation to one of the electricity market: although liberalized in principle, there are still regulated prices for private consumers. Full liberalization poses the same major challenges mentioned in the analysis of the electricity market.

664. Romania covers almost 70% of the domestic demand from its internal *production*. The two major producers are Romgaz (a state-owned company) and Petrom (a privately owned company). Additionally, Romgaz controls most of the gas storage facilities.

665. Production, transport and distribution are now vertically separated, with Transgaz acting as independent network operator of the transportation network. Tariffs for domestic transport activities are regulated. OPCOM operates the gas market, on which a significant number of independent suppliers are active, while distribution is handled by more than fifty regional companies. From 2007, the largest distributors have been obliged to separate their supply functions from the distribution ones.[880]

666. As was the case with the electricity market, the biggest challenge in the gas sector remains the regulated retail market. Similar to electricity, private consumers can opt for the liberalized market, but few do, given the fact that Romania has committed to fully liberalize the market only by 2018. The concern is the predicted price increase once domestic gas prices are aligned with international ones.

667. The Romanian government has been known to intervene in the gas market. For example, in 2011 it imposed a ban on domestic gas exports, during the financial crisis, which measure was repealed only in 2013, after an infringement procedure was launched by the European Commission.[881] Therefore, political constraints might determine the politicians in power to keep regulated prices in place as much as possible.

668. The regulator for gas is now the same as the one for electricity, NRAE. However, it has not been an independent body for long (until 2012 it acted under the coordination of the Prime Minister), and it does not command the same respect as other professional organizations such as the Competition Council. NRAE has no powers to enforce provisions of Competition Law, although it is tasked to ensure competitive energy markets. Thus, its functions overlap with those of the Competition Council. The cooperation between the two bodies has been formalized,

880. OECD, *Romania. Competition Law and Policy. A Peer Review* 81 (2014), available online at http://www.oecd.org/daf/competition/Romania-Competition-Law-Policy-2014-EN.pdf, last visited on 23 Mar. 2018.

881. Press Release: Free movement of goods: Commission requests ROMANIA to remove barriers to exports of natural gas, available online at: http://europa.eu/rapid/press-release_MEMO-14-470_en.htm, last visited on 23 Mar. 2018.

although, as seen in the case of electricity markets, it did not preclude the Competition Council from investigating the activity of NRAE for adopting acts which undermine competition for renewable energy. However, the close cooperation between the two is seen as necessary for the success of the transition to liberalized and competitive energy markets.[882]

669. To conclude, it is obvious that the energy sector poses great challenges to competition, although the normative documents and the market structure as well as the institutional framework are in line with EU requirements. The regulatory body is independent and vertical separation has been achieved by separating production, network operation, distribution and supply.

670. The obstacles faced by the energy markets stem from the fact that the government still controls almost all the electricity generation capacity and plays a strong role in natural gas production. In addition, price regulation continues to exist in both electricity and gas retail markets.

671. Unlike other European countries, Romania is determined to liberalize all its energy retail markets in the coming years. However, this will create major challenges: prices will increase; there will be more opportunities to invest in domestic and cross-border infrastructure, but it will be politically unattractive.[883]

672. With respect to the regulatory aspects related to the energy sector, a 2014 OECD report emphasized the need to continue focusing on electricity and gas markets, including strengthening of the necessary know-how and expertise within the Competition Council. This could only be achieved by a positive and productive cooperation with the regulatory authority within the energy sector (NRAE): the Council could assist NRAE in formulating policy proposals and regulations to end price regulation and could use its enforcement powers to prevent any conduct that would undermine the competitiveness of the energy markets.[884]

---

882. OECD, *Romania. Competition Law and Policy. A Peer Review* 83 (2014), available online at http://www.oecd.org/daf/competition/Romania-Competition-Law-Policy-2014-EN.pdf, last visited on 23 Mar. 2018.

883. OECD, *Romania. Competition Law and Policy. A Peer Review* 88–89 (2014), available online at http://www.oecd.org/daf/competition/Romania-Competition-Law-Policy-2014-EN.pdf, last visited on 23 Mar. 2018.

884. *Ibid.*, at 94.

Conclusions

673. This monograph has examined and explained various aspects of Romanian energy law and energy markets. It covered regulatory aspects concerning electricity, natural gas, coal and oil. It also touched upon environmental, tax and competition law where they interacted with energy law. The purpose was to provide the reader not only with an integrated and complete image of energy regulatory framework in Romania, but to underline the importance of the field and the relations between the energy sector and other various parts of law.

674. These conclusions wish to highlight some of the peculiarities of the Romanian energy markets and Romanian energy law. Obviously, the two are strongly intertwined for the provisions of the legislation will strongly influence the functioning of the markets, but at the same time, the peculiarities of the market might cause or call for changes in legislation. Therefore, a single conclusion, which would fit all energy markets and topics covered here, is very hard to draw.

675. There are two central topics heavily debated in Romania: on the one hand, the liberalization of the energy markets for electricity and gas, and, on the other hand, the diversification of energy sources. However, the political divergences are more related to the calendar and methods of implementation, than to different options or strategies.

676. The liberalization of the energy market was mandatory once Romania joined the EU, but even today Romania lags behind. As indicated in the first part, the Romanian electricity market is not yet fully liberalized, with the exception of the wholesale market. The retail market is fully opened since 2007 for business clients who wish to change their current supplier and its liberalization should have been complete by the end of 2014. However, natural persons still enjoy the benefits of a regulated retail market, although the government has promised to liberalize this market as well until the end of 2018.

677. Notwithstanding all the above, the problem is far from being solved. The current framework refers to vulnerable consumers, but failed to define them properly. This will affect the social security measures meant to protect a social category estimated at 1.5 million people. The remaining 7 million residential customers will have to adjust to the increase of prices, which may have significant political costs and a strong impact on the purchasing power of the population.

678. Things are not much different with regards to the gas market. The retail market is still regulated, but according to the new calendar it will be fully liberalized only in 2018. The new rules will affect around 2.8 million residential customers and more than 170,000 industrial customers. However, if the decision to postpone the liberalization of the retail market could be explained by the social effects which will be caused, allowing large industrial producers to still benefit from regulated prices is hard to understand. Supporters of liberalization claim that these companies are political clients and thus they are allowed to continue to function despite their lack of economic efficiency. The opponents claim that complete liberalization would have an undesired impact on the competitiveness of Romanian products and might lead to closure of businesses and massive layoffs.

679. According to the Competition Council, the unbundling required by the community regulations was successful and the market is fairly competitive. However, cross-ownership is still a common phenomenon. There are still lots of challenges for the regulators and the enforcing agency.

680. Romania is following the new trend of concern for the climate and environment and is taking big steps towards the transition from traditional fossil fuels to renewable energy sources. However, a sustainable energy mix still has to achieved, even if Romania has reached in 2014 its target for renewable energy production which was set for 2020.

681. Last but not least, it is estimated that Romania's subsoil holds oil reserves in the value of USD 200 billion. However, the recovery of these reserves requires strengthening the competition on the market, which is 90% controlled by Petrom and Romgaz and by increasing both investments in new technologies and production costs. For the moment, Romania remains a medium-size producer, but still in the top ten producers of Europe and Asia. Smaller players have started entering the market, but their presence remains minor. Thus, it is hard to estimate a significant change in the status quo of the Romanian oil market in the near future.

Selected Bibliography

Romania's Energy Strategy for 2007–2020, http://www.minind.ro/energie/STRATEGIA_energetica_actualizata.pdf.

Adrian Mosoianu. Romania va importa luna viitoare, ianuarie 2015, doar 2% din necesarul sau de consum de gaze naturale, Energy Report, 23.12.2014, available online, in Romanian only, at: http://www.energyreport.ro/index.php/2013-petrol-si-gaze/2013-stiri-petrol-si-gaze/2013-rafinare-si-marketing/3401-romania-va-importa-luna-viitoare-ianuarie-2015-doar-2-din-necesarul-sau-de-consum-de-gaze-naturale.

Alexandra Pele – Harta intereselor energetice ale americanilor si rusilor in Romania, Gandul.info, 18.10.2013, available online in Romanian only at: http://www.gandul.info/financiar/harta-intereselor-energetice-ale-americanilor-si-rusilor-in-romania-11528530.

AOAR: Noua lege a energiei permite instaurarea moopolului, 19.9.2012, available online in Romanian only at: http://www.business24.ro/energie/stiri-energie/aoar-noua-lege-a-anergiei-permite-instaurarea-monopolului-1519092.

Aurel Dragan. Dupa un secol de exploatare, mai avem gaze doar pentru 10 ani, Capital.ro, 30.11.2013, available online, in Romanian only, at: http://www.capital.ro/189384.html.

Catalin Stanescu. 'Romania: The Target for Renewable Energy Production for 2020 Has Already Been Reached – but Is It Really Good News?'. *Renewable Energy Law and Policy Review* 6, no. 1 (2015).

Centrul de Politici Durabile. Analiza taxelor de mediu in Romania, pp. 18–19, available online in Romanian only at: http://www.ecopolis.org.ro/files/Analiza%20taxelor%20de%20mediu%20in%20Romania.pdf.

Colectiv. Noul Cod civil – comentarii, doctrina, jurisprudenta – Vol I (The New Civil Code – Comments, Doctrine and Case Law – Vol I), Ed. Hamangiu, Bucuresti, 2012.

Consiliul Concurentei. Directia Industrie si Energie – Raport privind rezultatele preliminare ale investigatiei sectoriale pe piata energiei electrice din Romania, available online, in Romanian only, at http://www.consiliulconcurentei.ro/uploads/docs/items/id10016/raport_investigatie_utila_energie_electrica_07012014-forma_neconfidentiala-pub.pdf.

Consiliul Concurentei. Raport Anual 2014, p. 4, available on line, in Romanian, at: http://www.consiliulconcurentei.ro/uploads/docs/items/id10061/raport_anual_2014.pdf.

Selected Bibliography

Daniel Yergin. *The Prize. The Epic Quest for Oil, Money and Power*. New York: Simon & Schuster, 1991.

Economic Commission for Europe, Committee on Environmental Policy – Environmental Performance Reviews: Romania, published by the United Nations, 2001, available online at: http://www.unece.org/fileadmin/DAM/env/epr/epr_studies/romania.pdf.

Everett Rosenfeld. 'European Shale Dream Is Dying before it started'. 23.2.2015, cnbc, available online at http://www.cnbc.com/id/102446971.

Harta Resurselor Romaniei. In cati ani se vor termina rezervele de gaze natural, petrol si carbune, 30.12.2013, gandul.info, available online, in Romanian only, at: http://www.gandul.info/financiar/harta-resurselor-romaniei-in-cati-ani-se-vor-termina-rezervele-de-gaze-naturale-petrol-si-carbune-11748644.

Nadine Bret-Rouzaut & Jean-Pierre Favennec. *Oil and Gas Exploration and Production. Reserves, Costs, Contracts*. 3rd edn. Paris: Editions Technip, 2011.

OECD – Romania. Competition Law and Policy. A Peer Review. (2014): 76, available online at http://www.oecd.org/daf/competition/Romania-Competition-Law-Policy-2014-EN.pdf.

Romulus Cristea. Descoperire istorica de gaz in Marea Neagra. Ce are de castigat Romania, 22.02.2012, Romania libera, available online, in Romanian only, at http://www.romanialibera.ro/actualitate/eveniment/descoperire-istorica-de-gaz-in-marea-neagra--ce-are-de-castigat-romania-254582.

Roxana Ionescu. 'An Overview on Environmental Law in Romania'. *PLC Cross-border Environment Handboo,* 2008–2009, available online at http://www.nndkp.ro/publications/articles/environment-handbook_.

Valeriu Stoica. 'Drept Civil'. *Drepturile Reale Principale* 1, (Civil Law. Main Real Rights. Vol. 1), edited by Humanitas, Bucuresti, 2004.

Index

The numbers here refer to paragraph numbers.

Balancing market operator, 76, 170, 297
Balancing market, 76, 133, 135, 137, 170, 272, 273, 292, 297
Bio-diversity, 534
Bilateral contracts market, 180, 184, 186–193
Biomass, 109, 110, 186, 612
Biofuels, 117, 612
Biogas, 109, 237, 265, 295

Challenge (administrative), 6, 62, 83, 360–364, 490, 505, 643, 654, 663, 666, 669, 671, 679
Cross-subsidy, 112
Climate change, 531, 578–580
Coal, 440–529, 587, 604, 607, 608, 612, 673
Cogeneration, 57, 59, 70, 76, 77, 104, 107–116, 135, 186
Competition Council, 70, 72, 142, 179, 258, 300, 634, 635, 638, 639–643, 647, 650, 653-659, 661, 668, 672, 679
Competition Law, 37, 636–639, 641, 642, 643, 648, 661, 662, 668, 673

Dangerous substances, 466
Distribution System Operator, 162
Distribution system, 161–162, 230, 250, 280, 287, 288, 290, 291

Emissions, 63, 80, 111, 529, 578–580
Environmental impact assessment, 533, 563–575
Environmental permit, 536, 558–563
Export, 131, 170, 226, 269, 319, 436, 605, 653, 657, 659, 667
Excise tax, 586–588, 596, 599, 602–614

Farm-Out Agreement, 329, 368–408

Green Certificates, 73, 118, 121, 122, 124, 180–200, 648
Green Certificates market, 71, 121, 180–200, 648

Hydro-power, 109

Import, 117, 131, 140, 170, 225, 226, 269, 294, 314–316, 319, 433, 438, 439, 596, 607, 633
Interconnectivity, 62, 631, 645

Natural gas, 51, 220, 223–231, 236, 237, 246, 257–266, 268, 269, 272–275, 278, 280, 281, 283, 285–288, 290–301, 303, 305–316, 424, 587, 592, 595, 596, 606, 612, 613, 621–624, 630, 633, 635, 653, 670, 673
Nuclear, 547
NRAE, 52, 68–73, 76, 78, 82, 98, 104, 113, 115, 121, 124–130, 133, 134, 137, 140–142, 144, 148, 150, 153, 154, 158, 159, 162, 163, 165, 168, 171, 174, 175, 177–180, 183, 198, 199, 201–203, 205–208, 211, 213–219, 229–235, 237, 239–243, 245–251, 254–258, 264, 267, 269, 270, 274, 280, 282, 286–288, 291, 293, 295, 297, 299, 300, 303–309, 315 608, 622, 648, 650, 654, 659, 662, 668, 672
NAMR, 123, 319, 321, 324, 338, 339, 341, 343, 344, 346–351, 353–356, 360–363, 365, 370, 373, 379, 380, 382–384, 396, 399, 401–405, 410, 412–415, 423, 427, 428, 430, 433, 434, 440, 442–444, 446, 452, 455, 458, 461–465, 471, 473, 475,

476, 487, 490, 493, 498–500, 502, 503, 507, 508, 512, 513, 517, 520, 523, 525, 526

Oil (*see* Petroleum)
Oil refinery, 579
OPCOM, 121, 180–183, 188–190, 194, 199, 648, 651–653, 655, 657, 659, 665

Peat, 441, 520, 604
Petroleum, 317–439, 595

Renewable energy, 68, 71, 73, 108–124, 133, 135, 136, 150, 180, 181, 182, 184, 632, 668, 680

Relevant ministry, 53, 61, 63, 66–68, 70, 99, 100, 102, 123, 140, 177, 178, 225, 226–228, 281, 466

System operator (*see* transmission and system operator)

Transmission and system operator, 53, 68, 71, 97, 100, 105, 121, 125–142, 144, 145, 149, 166, 170, 171, 174, 177, 181, 203, 212, 219

Unbundling, 54, 302, 679

Vertically integrated undertaking, 143, 156, 157, 159, 302